T0305853

High-Power Laser Material Processing for Engineers

This book focuses on the mechanisms of how laser light is produced, guided, and focused for materials processing, and these are explained in an easy-to-understand language for practical use. It emphasizes a basic understanding of the principles necessary to run lasers in a safe and efficient way and provides information for quick access to laser materials processing for laser users. The book exhibits the following features:

- Provides simple explanations and descriptions of complex laser material interaction mechanisms to help readers understand relevant effects during laser beam irradiation of materials.
- Explains the main high-power laser materials processing methods, giving hints to get started with the processing and how to avoid imperfections.
- Focuses on high-power laser applications that are explained in an accessible, descriptive way with practical explanations and minimal formulas.
- Teaches how to measure laser beam characteristics and how to install and handle laser equipment correctly.
- Gives practical advice on typical equipment arrangements and parameter ranges.

This practical handbook serves as a guide for students studying production technologies to learn about laser processes, and for engineers who want to start working with laser processes safely and quickly.

Joerg Volpp is Associate Professor at University West in Trollhättan, Sweden and Department of Engineering Sciences and Mathematics at Luleå University of Technology, Sweden. He received a doctoral degree from the University of Bremen, Germany, focusing on laser beam welding of aluminum alloys and spatter and pore formation mechanisms. In addition to conducting scientific and industrial research on high-powered lasers, he also teaches laser-related courses.

High-Power Laser Material Processing for Engineers

Joerg Volpp

CRC Press
Taylor & Francis Group
Boca Raton New York London

CRC Press is an imprint of the
Taylor & Francis Group, an **informa** business

Designed cover image: © Joerg Volpp

First edition published 2025
by CRC Press
2385 NW Executive Center Drive, Suite 320, Boca Raton FL 33431

and by CRC Press
4 Park Square, Milton Park, Abingdon, Oxon, OX14 4RN

CRC Press is an imprint of Taylor & Francis Group, LLC

© 2025 Joerg Volpp

Reasonable efforts have been made to publish reliable data and information, but the author and publisher cannot assume responsibility for the validity of all materials or the consequences of their use. The authors and publishers have attempted to trace the copyright holders of all material reproduced in this publication and apologize to copyright holders if permission to publish in this form has not been obtained. If any copyright material has not been acknowledged please write and let us know so we may rectify in any future reprint.

Except as permitted under U.S. Copyright Law, no part of this book may be reprinted, reproduced, transmitted, or utilized in any form by any electronic, mechanical, or other means, now known or hereafter invented, including photocopying, microfilming, and recording, or in any information storage or retrieval system, without written permission from the publishers.

For permission to photocopy or use material electronically from this work, access www.copyright.com or contact the Copyright Clearance Center, Inc. (CCC), 222 Rosewood Drive, Danvers, MA 01923, 978-750-8400. For works that are not available on CCC please contact mpkbookspermissions@tandf.co.uk

Trademark notice: Product or corporate names may be trademarks or registered trademarks and are used only for identification and explanation without intent to infringe.

ISBN: 978-1-0327-8189-1 (hbk)
ISBN: 978-1-0327-8191-4 (pbk)
ISBN: 978-1-0034-8665-7 (ebk)

DOI: 10.1201/9781003486657

Typeset in Times LT Std
by KnowledgeWorks Global Ltd.

Contents

Preface..vii
About the Author ix

Chapter 1   Introduction: The Laser Story in Short 1

Chapter 2   Theoretical Background 2

    2.1   Laser Theory 2
        2.1.1   What Is Light?...2
        2.1.2   Laser Principle 2
        2.1.3   Laser Active Medium 3
        2.1.4   Laser Resonator 5
    2.2   Laser Machines 7
        2.2.1   Laser Designs 7
        2.2.2 CO_2 Lasers...9
        2.2.3   Solid-State Lasers...11
        2.2.4   Diode Laser 17
    2.3   Characteristics of Laser Beam 18
    2.4   Beam Guiding and Beam Shaping 22
        2.4.1   Fiber Optics for Beam Guiding.......................23
        2.4.2   Beam Shaping Elements..................................25
        2.4.3   Laser Beam Focusing 28
        2.4.4   Laser Beam Shaping..30
    2.5   Laser Beam Energy Transfer...34
        2.5.1   Phenomenological Overview...........................34
        2.5.2   Absorption of Metals......................................36
        2.5.3   Heat Transfer in Materials..............................40
    2.6   Material Properties 41
        2.6.1   Surface Tension 43
        2.6.2   Phase Changes...43
        2.6.3   Diffusion...45
    2.7   Process Monitoring 46
        2.7.1   Process Emissions 46
        2.7.2   Process Sensors 47

Chapter 3   Laser Material Processes..51

    3.1   Overview 51
    3.2   Safety..53
    3.3   Surface Hardening...54
        3.3.1   Overview 54
        3.3.2   Martensite Hardening......................................55

		3.3.3	Laser Martensite Hardening	56
		3.3.4	Precipitation Hardening	58
		3.3.5	Beam Shaping for Surface Treatment	59
		3.3.6	Practical Issues	59
		3.3.7	Laser Shock Peening (for Hardening)	60
	3.4	Laser Forming		60
	3.5	Laser Beam Marking		62
	3.6	Laser Beam Brazing		64
	3.7	Laser Beam Welding		66
		3.7.1	Heat Conduction Welding	66
		3.7.2	Laser Deep Penetration Welding	71
		3.7.3	Weldability	85
		3.7.4	Laser-Hybrid Welding	92
		3.7.5	Imperfections and Defects	94
	3.8	Laser Beam Cutting		105
		3.8.1	Introduction	105
		3.8.2	Process Principle	106
		3.8.3	Process Modes	108
		3.8.4	Process Parameters	108
		3.8.5	Quality Criteria and Imperfections	110
	3.9	Laser Drilling		111
	3.10	Laser Alloying and Dispersing		115
	3.11	Laser Cladding		116
	3.12	Laser Beam Additive Manufacturing of Metals		119
		3.12.1	Principles of Laser Additive Manufacturing	122
		3.12.2	Directed Energy Deposition	123
		3.12.3	Laser Powder Bed Fusion	133
		3.12.4	Powder Sheet Additive Manufacturing	138
		3.12.5	Part Disassembling Using Laser Beams	139
		3.12.6	When to Use AM?	140
	3.13	Research Topics		142
		3.13.1	Laser Developments	142
		3.13.2	Laser Processes	143
	3.14	Summary and Outlook		148
		Note		149
Chapter 4	Symbols and Abbreviations			150
	Symbols			150
	Abbreviations			152
References				155
Index				165

Preface

The book is intended to create a handbook for engineers who want to quickly access the field of laser materials processing. It is addressed to engineering students as support for university lectures and also for those who want to develop skills in laser processing.

The book mainly focusses on the basic principles of laser machines and laser processes; however it does not cover all circumstances or all material specialties that might occur. The intention is to give an idea of general relations and principles.

Preface

About the Author

Joerg Volpp studied Mechanical Engineering at the University of Stuttgart (Germany), with a focus on control technique and laser material processing. He finished his doctoral thesis at the University Bremen (Germany) in the topic of spatter and pore formation during laser beam welding of aluminum alloys. He was promoted to docent and associate professor at Lulea University of Technology (Sweden) expanding the research topics to laser beam hardening, cutting, and additive manufacturing. Currently, he works at University West (Sweden) as a professor in material processing.

1 Introduction
The Laser Story in Short

The success story of the laser started with the developed concept of stimulated emission by Albert Einstein in 1916 proving Planck's law of radiation. The underlying concept that an atom in an excited state can emit a photon when it encounters another photon made the laser process theoretically possible. The incoming photon just needs the same energy as the excited atom, and both photons travel further with the astonishing properties of having the same wavelength, direction, polarization, and phase. However, naturally, the stimulated emission is very rare, and it took until 1928 to experimentally confirm the theory. In the 1950s, researchers took ideas and developed a concept that realized the stimulated emission in a resonator system.

Maiman [Mai60] developed the very first MASER by using ruby as laser active medium. The MASER emitted pulsed light. In 1960, the first laser was built. Korad and Raytheon already promoted the first industrial laser in 1963, built for drilling processes [Kla84]. Then, the laser concept was quickly further developed and is nowadays used for many purposes. Laser light is able to transport information and energy, which enables applications in various fields, e.g., in data transfer in information technology, medical applications and, of course, in material processing. Its unique features enable it to reach peak powers in the petawatt range, short pulses in the pico- and femtosecond range and small dimensions in the nanometer range. Laser light shows high coherence and therefore high measurement resolution and a distinct color.

Within manufacturing, dieboard slotting was the first industrial application of laser cutting demonstrated in 1970 at William Thyne Ltd (Edinburgh, UK) using a BOC Falcon 200 W CO_2 laser. Five years later, Ford showed the feasibility of car body part cutting with a 400 W CO_2 laser, while in 1978, the first car with laser-cut parts was launched. Laser beam welding was shown by Martin Adams in 1970. At the Welding Institute, he presented welded sections made with a CO_2 laser beam in 1.5 mm thick metal [Ada70].

The laser became a success story and is great for many applications. In materials processing, the laser enabled processes that were not possible before. However, to be honest, the laser is not perfect for everything. Therefore, this book shall help engineers to get a quick understanding of the main features of lasers for industrial processing and the related material interactions to learn about the main possibilities, challenges, and limitations that can occur and how to deal with them. This book concentrates on laser-based manufacturing of mainly metals with current state-of-the-art solid-state lasers.

DOI: 10.1201/9781003486657-1

2 Theoretical Background

2.1 LASER THEORY

2.1.1 WHAT IS LIGHT?

Light is electromagnetic energy. It can be defined as an electromagnetic wave consisting of electric and magnetic fields that oscillate perpendicular to each other. Albert Einstein proposed that light consists of bundles of wave energy, named photons [Ein05], and received the Nobel prize for finding the photoelectrical effect. Besides the description of light as a wave or a photon, nowadays, the dualism is handled by calling it a quantum mechanical system. However, both descriptions can help to understand and explain the behavior of light in different circumstances. In the wave theory, light is described by its wavelength and polarization (orientation of the E- and B-vectors). Due to its wave character, light shows interference phenomenon. The energy that light can transport depends on the frequency or wavelength of light. The higher the frequency, the higher the energy (Figure 2.1). Following the particle-like nature of light, a photon has (almost) no mass, no charge, but transports energy (in Joule)

$$E = \frac{h \cdot c}{\lambda},$$

with h being Planck's constant (6.625×10^{-34} J·s), c the speed of light (2.998×10^8 m/s), and λ the wavelength of light (in meter). The speed of light is high but not infinite, and reaches in vacuum a speed of 2.998×10^8 m/s, which is a universal constant. The speed of light is slower when traveling through a medium (e.g., air: 0.03% slower or glass: 30% slower). A light beam is considered a stream of photons.

Photons can interact with other particles such as electrons. The phenomena involving typical interactions of photons with a medium include absorption (transfer of photonic energy to another particle), reflection (path change of photons at a material interface), and transmission (passing of photons through the medium) that includes refraction effects.

Typical values to describe characteristics of a light stream include transported energy (in Joule), power P (in Watt) or flux (in Joule per second), flux density or power density or irradiance per unit area (in Watt per square meter or Joule per square meter, respectively).

2.1.2 LASER PRINCIPLE

The word laser is an acronym that stands for "Light Amplification by Stimulated Emission of Radiation". Nowadays, the word laser is used for the physical process of laser light creation as well as for the machine that produces the laser light. According to the standard ISO11145, in a laser machine, the laser light is produced inside a

DOI: 10.1201/9781003486657-2

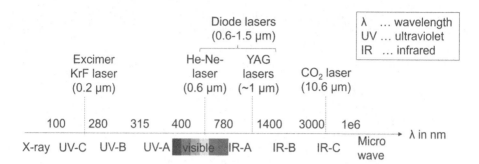

FIGURE 2.1   Wavelengths of light and wavelengths of selected laser systems.

resonator, where the laser active medium (LAM) is positioned (Figure 2.2). Energy input is needed for so-called pumping, which increases the energy state of the LAM. Most laser systems require active or passive cooling devices due to energy losses during the laser beam creation.

2.1.3   LASER ACTIVE MEDIUM

Three different mechanisms can occur during the interaction of light with material (Figure 2.3). When a photon interacts with an atom or molecule, it can absorb its energy and increase thereby the energy state of the atom or molecule (excitation of matter). This process is called absorption. According to Planck, photons have quantized energy states, which means that the energy transfer is a multiple of the minimum quantized energy.

Since most attenuated states are unstable, after a certain time, the atom or molecule prefers to go back to a more stable condition. This energy decrease can then lead to a so-called spontaneous emission of radiation. The exact time and direction of the emission can be predicted by statistical methods but are not precisely defined. Absorption and spontaneous emission are natural mechanisms that can be observed all the time.

However, stimulated emission is a special case that does not occur regularly in nature but is necessary for laser light production. It is possible that a photon can stimulate the energy decrease of an excited atom or molecule. Thereby, the energy release can be transferred to produce another photon (similar to the spontaneous

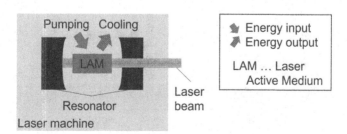

FIGURE 2.2   Systematic sketch of the main components needed to build a laser machine.

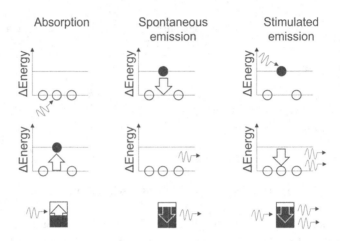

FIGURE 2.3 Principles of light-matter interaction.

emission), called relaxation of light. When this happens, the stimulating photon and the newly created photon have identical characteristics. They are exiting the system with the same wavelength, direction, polarization, and phase. Since there is one photon entering the active atom/molecule of the LAM and two photons are exiting, this process can be seen as light amplification.

In nature, spontaneous emission occurs more likely than stimulated emission. According to Boltzmann's law, the population of energy states in a system shows less population in higher states (Figure 2.4). However, for a stimulated emission, the so-called population inversion is necessary, which means that in a higher energy state, a higher population must be present. Since this state is not possible in thermodynamic equilibrium, population inversion must be created to enable laser light production. This requires thermal non-equilibrium.

A pre-requisite for laser light production is the maintenance of population inversion for a certain time. Therefore, the LAM needs to be pumped to enable

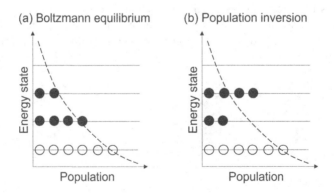

FIGURE 2.4 Principle sketch of population distributions at (a) thermodynamical equilibrium (Boltzmann equilibrium) and (b) population inversion.

the necessary energy input. Pumping describes the energy input into the LAM to increase its energy state, which later enables population inversion, having more population at a higher energy level. At the same time, losses occur, which lead typically to the requirement to have a cooling system.

Energy input can be achieved in different ways depending on the nature of the LAM; typical are optical pumping (absorption), electrical discharge (electron collisions), electric processes (semiconductors), or chemical processes. Most typical LAMs for lasers in production are solids and gases.

Solid-state lasers use a host medium, typically a YAG (Yttrium-Aluminum-Garnet) or SiO_2 crystal, in which the LAM is dotted. Neodymium ions (Nd^{3+}) were found to be feasible material for laser systems; however, ytterbium (Yb), being a more energy-efficient material, started to replace it. Independent from the used design (rod, disk, fiber), those systems are usually optically pumped by high-power flashlights or nowadays, by diode laser light.

The most typical gas laser in production is the CO_2 laser. The CO_2 gas is excited by electrical discharge, while first the added nitrogen (N_2) is excited, which transfers energy to the CO_2 molecules that enables stimulated emission.

2.1.4 LASER RESONATOR

The sole energetic pumping of LAM would not lead to a directed laser beam yet. Therefore, the LAM is placed inside a resonator (Figure 2.5). The resonator enables multiple paths of photons by reflections through the LAM to increase amplification of

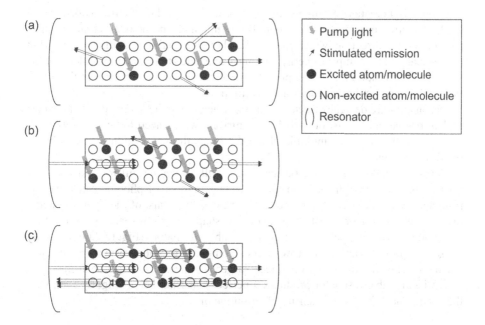

FIGURE 2.5 Principle sketch of the LAM in the resonator at (a) starting of pumping, (b) amplification initiation, and (c) stable amplification.

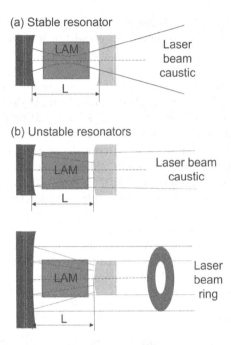

FIGURE 2.6 Sketches of typical resonator designs with spherical mirrors of (a) a stable resonator and (b) unstable resonators.

laser light and create a standing wave. The photons guided by the resonator through the LAM enforce stimulated emission. Both photons from the interaction are emitted in the same direction, which enables the photons to build a standing wave within the resonator. The pumping energy continues to excite atoms and molecules, but the stimulated emission in a perfect process continuously emits light only in the direction of the photons that already travel within the resonator.

Resonators are designed to enable the laser beam to exit at some point. The most typical resonators consist of two-shaped mirrors, which need to be precisely positioned at a distance L as multiple of the 0.5 times the wavelength λ to enable the build-up of a standing wave (Figure 2.6).

Stable resonators (Figure 2.6a) can use a fully reflective mirror and one mirror that is partly transmissive so that a certain number of photons can leave the resonator. Since the laser beam shows a caustic with a waist of the radius w_0 and a divergence, the mirrors must show a concave shape. Stable resonators are used for continuous wave (cw) solid-state lasers. Unstable resonators (Figure 2.6b) can only operate in pulsed width mode (pw lasers). Compared to stable resonators, unstable resonators can emit higher power since a larger volume of LAM can be used.

An ideal stable resonator produces a perfect Gaussian beam (Figure 2.7a), with the spatial intensity (power density) distribution of

$$I(r) = \frac{2 \cdot P_L}{(w^2 \cdot \pi)} \cdot e^{-\frac{2 \cdot r^2}{w^2}},$$

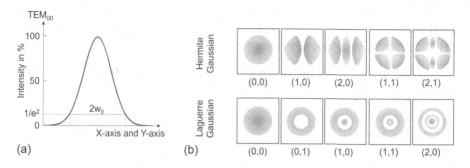

FIGURE 2.7 Spatial intensity distributions of laser beams showing (a) a perfect Gaussian beam and (b) Hermite-Gaussian and Laguerre-Gaussian beam profiles.

with the laser power P_L, beam waist w, and radius r. Further solutions of the wave equation can be derived leading to possible modes of the laser beam that can start vibrating in the resonator (Figure 2.7b).

2.2 LASER MACHINES

2.2.1 LASER DESIGNS

Different laser designs were developed with different LAMs, different principles of pumping, cooling, and resonator designs. The main aim is always to efficiently provide energy to LAM to enable population inversion and emit laser light with the lowest possible heat losses. Several laser designs are commercially available. Typical industrial high-power systems are CO_2 lasers, Nd:YAG rod lasers, Yb:YAG disk and fiber lasers, and direct diode lasers. CO_2 lasers are widely used, e.g., for cutting processes. Compared to gas lasers, like the CO_2 laser (wavelength 10.6 µm), the solid-state lasers (rod, disk, and fiber) emit at a wavelength around 1 µm and show therefore different caustic characteristics (Figure 2.8). Diode lasers at different wavelengths are also available for high-power applications. Typical values of selected high-power laser systems are given in Table 2.1. A smaller beam parameter product (BPP) indicates a better focusability of the laser beam.

In 2017, the total laser sales were $12.3 billion with increasing tendency for all laser types. The main markets are material processing and communication (each 1/3), R&D military, medical, lithography, instrumentation sensors, displays, optical storage, and printing [Las18].

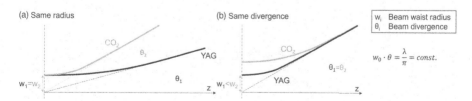

FIGURE 2.8 Influence of the wavelength on the beam caustic at constant BPP, showing (a) higher divergence and (b) larger beam waist of CO_2 laser beams.

TABLE 2.1
Typical Properties of Laser Systems

	CO_2	Diode Pumped Nd:YAG	Yb:YAG Disc	Yb:YAG Fiber
Typical beam parameter product (BPP) (mm*mrad)	5	12	10	12
Typical fiber diameter (μm)	–	400–1000	200	100–200
Typical maximum output power (kW)	20	6	8	20
Power efficiency (%)	5–8	10–20	10–20	20–30

Besides BPP and beam quality M^2, the efficiency of systems is used for the comparison of laser systems. The wall plug efficiency η_{WP} describes the total energy loss from the electrical plug power input P_{total} to the laser output power P_L.

$$\eta_{WP} = \frac{P_L}{P_{total}}.$$

Energy losses occur due to electrical transfer of energy to the pump source and the transformation of energy to be available for pumping, e.g., from electrical energy to light. Since not all the energy from pumping can be transferred to LAM, losses occur during this transfer as well (Figure 2.9).

In the LAM, quantum losses are systematic losses that appear since not all energy input from the pumping can be transferred to the photon energy due to the energetic scheme of LAM. The quantum efficiency η_Q is defined as the energy input into the produced photon (energy difference between the upper, u and lower, l laser level)

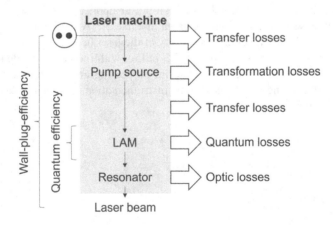

FIGURE 2.9 Overview of typical energy losses from the electric plug to the laser beam output.

$\Delta E_{u,l}$ in relation to the energy need for pumping E_{pump} (energy difference between the base, b, and excited, e, level) $\Delta E_{e,b}$

$$\eta_Q = \frac{\Delta E_{u,l}}{\Delta E_{e,b}} = \frac{h \cdot f}{E_{pump}}$$

Further losses occur in optical systems. All losses are transferred into heat, which requires cooling of the components to avoid thermally induced damages.

2.2.2 CO_2 LASERS

The most industrially used laser type is the CO_2 laser, which is a representative of gas lasers. Other gas lasers use Argon ions or Helium-Neon (He-Ne) gas mixtures. He-Ne laser is often found in industrial machines as pilot laser that is guided through the optical system to support the alignment and positioning of the laser beam in relation to the work piece.

As indicated by the name of the laser, the LAM is a gas. For CO_2 lasers, the energy increase is induced by the excitation of CO_2 molecules into different vibration states, which are quantized. This means that only certain energy states can be reached. Figure 2.10 shows the typical excitation states of CO_2 molecules in a CO_2 laser system. A specialty in CO_2 lasers is the use of N_2 molecules that are excited first by electric glow discharge pumping before they transfer the gained energy to CO_2 molecules.

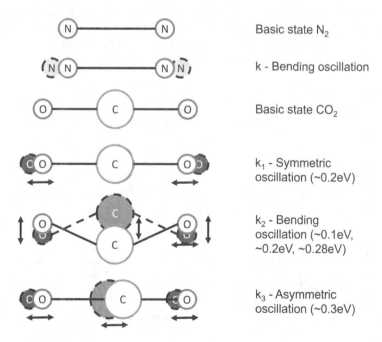

FIGURE 2.10   Vibration states of N_2 and CO_2 molecules.

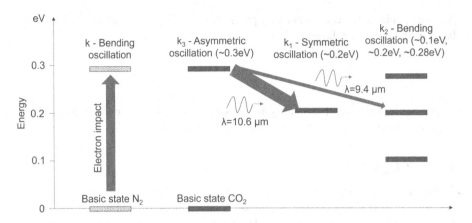

FIGURE 2.11   Energy levels of N_2 and CO_2 molecules, including excitation and relaxation states with photon emission relaxations.

The achievable energy levels and related vibration states are shown in Figure 2.11. Changes in the energy levels of CO_2 initiate the creation of laser light. The main wavelength created during de-excitation is 10.6 μm, while a small number of photons emit at 9.4 μm as well.

The electrical discharge pumping initiates N_2 excitation (Figure 2.12), while the energy is transferred to the CO_2 molecule, which is the upper laser level (state 4). Relaxation to state 3 (lower laser level) initiates the creation of photons. Further de-excitation to state 2 and the ground level (state 1) mainly results in energy transformed into heat. In order to keep population inversion (more molecules in state 4 compared to state 3), the de-excitation of the lower laser level (state 3) is supported by He atoms that are added to the gas mix in a CO_2 laser.

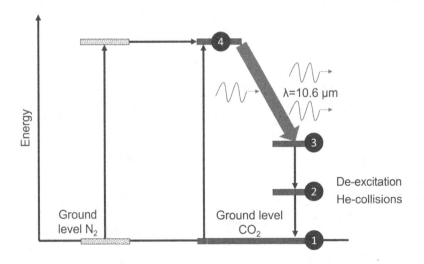

FIGURE 2.12   Schematic energy levels in a CO_2 laser system.

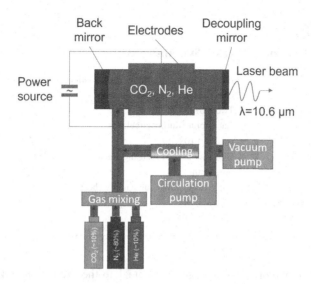

FIGURE 2.13   Sketch of a CO_2 laser system design.

As can be seen in Figure 2.12, the energy needed to pump the system is larger than the energy transferred to the photon of the laser beam. Therefore, a systematic energy loss occurs. The so-called quantum efficiency of typical CO_2 laser systems is 10–15%, which means that only 10–15% of the input energy can be transferred to the laser photons through pumping. Typically, active cooling of optical elements and gas mixture is therefore necessary.

A schematic sketch of a typical CO_2 laser system is shown in Figure 2.13. Pumping is done by electric glow discharge by a high-power, high-frequency power source through electrodes in the gas-loaded system, which shows the typical purple glowing. The LAM is CO_2, while N_2 and He gases are further added into the system. N_2 is needed for excitation and He for relaxation. The laser beam is established in the gas system within the resonator consisting of a non-transmissive back mirror and a partly transmissive decoupling mirror, wherefrom the laser beam exits the system. Cooling of gases is done by circulating the gas mixture and implemented cooling devices.

Typically, BPP > 3.4 mm*mrad (minimum diffraction limit for a certain wavelength) is achievable. The main characteristics of CO_2 laser systems are summarized in Table 2.2.

For safety issues, eyes must be covered by protection glasses with the appropriate filtering of wavelength. The laser light is not visible to human eyes.

2.2.3   SOLID-STATE LASERS

Solid-state lasers use solid material as LAM, which is made up of amorphous glass or dielectric crystals with dotted active material. The active ions are of rare earths or transition ions that have discrete electronic levels (Figure 2.14). Pumping is done

TABLE 2.2
Typical Data of CO_2 Laser Systems

Pumping	Electrical glow discharge: N_2 excitation, energy transfer to CO_2
LAM	CO_2 change of vibration states
Resonator	Two mirrors
Cooling	Gas cooling during circulation
Wavelength	(mainly) 10.6 μm
Beam quality	High, but limited focusing due to high wavelength
Beam transport	Free beam
Invest	Large devices, but comparably cheap
Wall plug efficiency	~5–15%
Output power	~20–100 kW

by high-energy photons from the ground state 1 to the excited state 4, lifting the electrons of ions into higher energetic states. De-excitation into the upper laser level (state 3) occurs that further continues to the lower laser level (state 2) to create laser photons, followed by further relaxation into the ground level.

In order to create population inversion, the upper laser level (state 3) must be constantly and reliably filled with excited ions, while state 2 should have a low population. The used dotted ions show the necessary characteristics to enable these characteristics. The transfer of ions from level 4 to level 3 is very fast that fills level 3, while the ions remain in level 3 for a comparably long time. The transfer of ions from level 2 to level 1 is also fast, which enables quick emptying of level 2. Typical

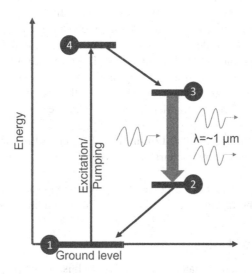

FIGURE 2.14 Sketch of the energy level system in a four-level LAM.

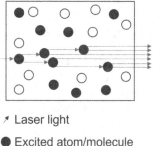

↗ Laser light

● Excited atom/molecule

○ Non-excited atom/molecule

FIGURE 2.15   Laser light creation in a solid-state LAM using a crystal with dotted material.

quantum efficiencies of those systems are around 50%. Optical pumping is used to excite ions using a light source. The laser process in the LAM is initiated and developed as seen in Figure 2.15.

Compared to other systems, solid-state lasers show relatively long lifetimes of the excited states, which enables higher energy storage and increased efficiencies.

2.2.3.1   Nd:YAG Rod Laser

Typical materials used in industrial laser systems are Nd:YAG crystals, which contain Nd^{3+} embedded in a YAG crystal. A rod laser uses a crystal rod that contains LAM (Figure 2.16), which is pumped by a flashbulb or by laser diodes. The resonator consists of two mirrors—one mirror is highly reflective and the outcoupling mirror is partly transmissive. Both the pump light source and the LAM typically need to be cooled. The possible output power of one rod is around 500 W, while a serial arrangement of rods can increase the output power of the laser system. The produced laser light has the wavelength of 1.064 µm. Laser light at that wavelength enables fiber guidance.

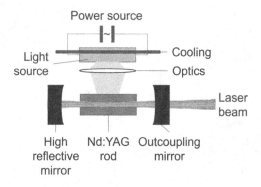

FIGURE 2.16   Principle sketch of a solid-state laser setup.

TABLE 2.3
Typical Data of Nd:YAG Rod Laser Systems

Pumping	Optical by flashbulb or diode lasers
LAM	Excited Nd^{3+} in YAG crystal rod
Resonator	Two mirrors
Cooling	Water or air cooling
Wavelength	1.064 μm
Beam quality	Comparably low
Beam transport	Fiber guided
Invest	Compact devices, but comparably expensive
Wall plug efficiency	~2–4% (flashbulb-pumped), ~10% (diode-pumped)
Output power	Typically ~2–10 kW

Typical data about Nd:YAG laser systems are summarized in Table 2.3.

2.2.3.2 Yb:YAG Disk Laser

New designs of solid-state lasers were developed in order to improve cooling possibilities and limit thermal lensing effects. A rod crystal with laser beam oscillating through the center can show steep thermal gradients. When laser light is produced, the temperature in the center of the rod is high compared to the circumference where typically water cooling is applied (Figure 2.17a). This temperature difference leads to inhomogeneous deformation of the rod due to thermal expansion and bulging

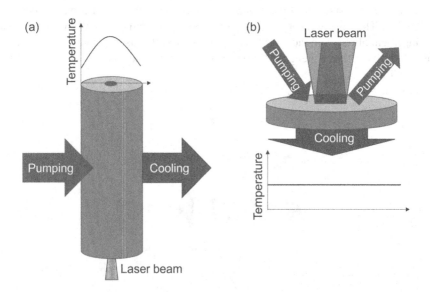

FIGURE 2.17 Schematic sketch of laser crystals and temperature distributions in (a) a rod and (b) a disk.

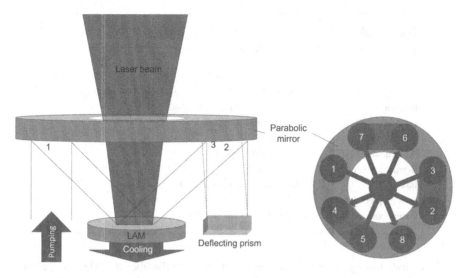

FIGURE 2.18   Sketch of a disk laser design.

of rod surfaces (thermal lensing effect). In combination with the temperature-dependent change of the refractive index, defocusing of the laser beam can occur at the end of the rods, which can lead to a misalignment or loss of laser light. In order to enable homogeneous cooling, disk laser systems have been developed [Gie05], where the rod is reduced to a thin disk that can be homogeneously cooled from the rear (Figure 2.17b).

However, by shrinking the rod to a disk, the distance of the laser beam propagating through the crystal is reduced in this design, which means that the amplification is reduced. Therefore, a smart laser design has been developed that leads the pump laser beam multiple times through the crystal. In order to guide the pump light, several prisms and parabolic mirrors are necessary (Figure 2.18). The rear side of the disk is used as a mirror (part of the resonator) for pump light and laser light.

Typically, ytterbium ions (Yb^{+3}) are used as the dotted material for disk laser in YAG crystals (LAM). Optical pumping with laser diode light enables a high-energy transfer efficiency due to the possibility to pump with the correct wavelength that can be actually absorbed by the LAM, which is 941 nm for Yb^{+3}. This enables ~90% absorption of the pump energy into the laser active material. The quasi-3-niveau system has a lower laser level already filled with 7.6% at 100°C due to being close to the ground level. The main data about disk lasers are summarized in Table 2.4.

2.2.3.3   Fiber Laser

A second design was developed based on the rod design to increase cooling possibilities and avoid thermal lensing effect in solid-state lasers. Instead of reducing the rod shape to a disk, long fibers dotted with typically Yb^{+3} are used. The generated large circumferential surface of the fiber is easy to cool, while guaranteeing a long path of the laser beam inside the LAM (Figure 2.19).

TABLE 2.4
Typical Data of Yb:YAG Disk Laser Systems

Pumping	Optical diode lasers
LAM	Excited Yb^{+3} in YAG crystal disk (quasi-3-niveau system)
Resonator	Mirror and rear part of the disk
Cooling	Water cooling of the disk
Wavelength	1.030 μm
Beam quality	Comparably high
Beam transport	Fiber guided
Invest	Compact devices, but comparably expensive
Wall plug efficiency	>25%
Output power	Scalable at ~5 kW per disk (cw)

The diode laser generated pump light is guided into the fiber core cladding, where the pump light is lead through the fiber with multiple reflections at the fiber cladding. Asymmetric fiber cores can improve the energy coupling from the pump light into the LAM. Double-cladding fibers can help to avoid the necessity of high-quality pump light. The pump light initiates the laser process in the fiber core (LAM). Mirrors at the ends of the fiber form the resonator. Fiber lasers show a high gain and high efficiency. The main features of fiber lasers are summarized in Table 2.5.

2.2.3.4　Pulsed Lasers

Several possibilities exist to pulse a laser machine. Intrinsic pulsing happens when the laser medium is not able to maintain population inversion due to the energetic state configuration of the LAM. This happens, e.g., in the first demonstrated Ruby Laser, where after laser emission, pumping has to regenerate population inversion again, since the upper laser level is emptied faster than the pumping can fill it.

A simple way of pulsing a laser beam is a mechanical shutter, e.g., a rotating disk with an opening at the outlet of the laser beam of the resonator. These lasers

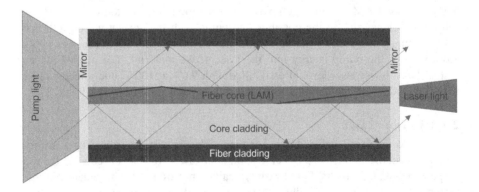

FIGURE 2.19　Schematic sketch of a fiber laser design.

TABLE 2.5
Typical Data of Yb:YAG Fiber Laser Systems

Pumping	Optical diode lasers
LAM	Excited Yb^{+3} in YAG crystal (quasi-3-niveau system)
Resonator	Mirrors at the fiber ends
Cooling	Water cooling of the fiber
Wavelength	1.070 μm
Beam quality	Comparably very high
Beam transport	Fiber guided
Invest	Compact devices, comparably not so expensive
Wall plug efficiency	>25%
Output power	Scalable at ~600 W per fiber unit (cw)

are called quasi-cw-lasers (QCW). However, the maximum energy peak is still the same as for a continuous emission, which means that the average energy emission is reduced compared to cw emission. The unused laser energy is transformed into heat on the rotating shutter when not emitted, which requires additional cooling.

In addition, emission control is possible by the input of the pumping source, which can be pulsing of the pump light. More efficient use of the population inversion is possible by creating laser pulses using quality-switching (Q-switching). Optical pumping is used to increase population inversion, while the losses are kept at a high level. That way, no stimulated emission happens yet. Switching the laser on induces stimulated emission and a short pulse can be created. Active elements (e.g., electro-optical or acoustic-optical modulators) can create pulses in the nanosecond range, while passive elements enable shorter pulses. Such lasers are often used in marking systems or micromachining. Mode locking enables to synchronize modes in the resonator, which leads to interference. Constructive interferences enable high-peak short pulses. Pulse durations of pico- and femtoseconds are possible.

2.2.4 DIODE LASER

A diode laser uses semiconductors to produce laser light and has the advantage to directly create laser light from electricity without creating, e.g., pump light to transfer the energy into the LAM. Diode lasers are typically used as pump source for fiber or disk lasers, but are also increasingly used directly without further transfer for material processing.

The LAM is the p-n-transfer of the semiconductor, while pumping is directly enabled by electrical energy. The laser emission is induced when electrons pass the p-n-transition (Figure 2.20a). A single p-n-transfer induces about 5 mW to 50 mW laser energy output. A single diode emitter is limited to a few Watts energy emission and produces comparably divergent light.

A combination of multiple laser diodes assembled to bars increases the output power to, e.g., 60–120 W using 20–25 laser diodes (Figure 2.20b). The challenge of

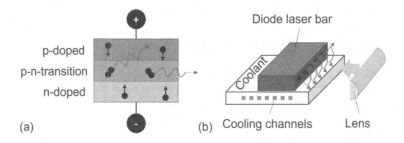

FIGURE 2.20 Principle sketches of (a) a laser diode p-n-transition and laser emission and (b) a diode laser bar.

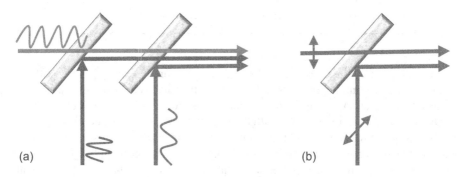

FIGURE 2.21 Principles of (a) wavelength coupling and (b) polarization coupling using optical elements.

emitted diode laser light is the divergence of the laser beam that requires typically the installation of micro-lenses to enable a more focused laser beam. The so-called stacks of multiple diode bars are typically used to further increase the laser output power. Wavelength coupling and polarization coupling result in increasing the laser power even more by combining several diode laser beams (Figure 2.21).

2.3 CHARACTERISTICS OF LASER BEAM

Laser beams have unique characteristics that distinguish laser light from other light sources (Figure 2.22). Due to the creation process of laser light, all photons in a laser

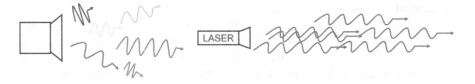

FIGURE 2.22 Comparison of light emitting from a conventional light source and a laser.

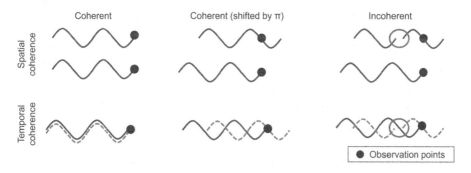

FIGURE 2.23 Spatial and temporal coherence of light waves.

beam have the same energetic state, which means that they have the same wavelength. Therefore, laser light has in principle only one color (monochromatic). Laser light waves have a strong phase correlation and are directed in one direction (with a certain but small divergence angle compared to other light sources).

Laser light has a high coherence in time and space, which means that two wave fields have a defined phase shift for a long travel distance. This property enables the laser beam to show interference effects. Spatial coherence can be measured by the double-slit-test, while temporal coherence can be measured in, e.g., the Mach–Zehnder interferometer. Considering two geometrical non-identical observation points in space, the spatial coherence is given when two parallel light waves show a constant phase shift (Figure 2.23). Whereas, temporal coherence compares the phase relation of two light waves observed in one observation point. The coherence is higher at longer times of a constant phase relation.

As indicated before, a laser beam is a stream of single photons and does not propagate perfectly parallel. A laser beam shows a caustic with a minimum diameter at the waist and a divergence θ along the propagation direction. The spatial beam radius is defined as the distance from the beam center where the intensity drops to $1/e^2$ ($\approx 13.5\%$) (Figure 2.7). This means that the beam area $\pi \cdot w^2$ contains 86.5% of the entire laser power. The more precise physical definition of the laser beam area uses the second moment approach (ISO 11146). A Gaussian beam propagates along the z-axis showing a planar wave front in the waist changing into a curved wave front (Figure 2.24).

A laser beam can be described using the two introduced characteristics, namely the beam waist radius w_0 and the divergence θ. Thereby, *BPP* is defined as

$$BPP = w_0 \cdot \theta = \text{constant.}$$

The *BPP* is constant for the whole beam propagation. A perfect Gaussian beam has, in theory, a *BPP* of

$$BPP = w_0 \cdot \theta = \frac{\lambda}{\pi}.$$

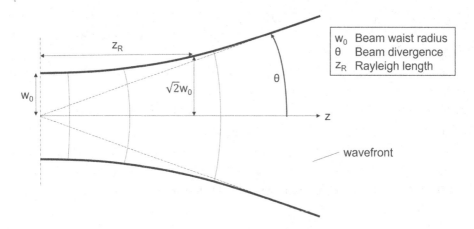

FIGURE 2.24 Laser beam propagation and caustic properties.

All laser beams in reality show a higher *BPP* due to losses during the laser creation process. This is considered by the beam quality parameter M^2

$$M^2 = \frac{\pi \cdot w_0 \cdot \theta}{\lambda}.$$

When describing oval laser beams, M^2 is defined in two different axes. M^2 has a theoretical minimum value of 1 for a perfect Gaussian beam.

Therefore, the beam waist radius of a real beam is defined as

$$w_0 \approx M^2 \frac{\lambda}{\pi} \cdot \frac{f}{w_{lens}}.$$

The general rotational-symmetric laser beam caustic radius $w_{(z)}$ along the propagation direction z, not limited to Gaussian beams, is

$$w_{(z)} = w_0 \cdot \sqrt{1 + \left(z / z_R\right)^2}.$$

The divergence can be derived including the laser beam radius w_L with

$$\theta = arctan\left(w_L/f\right)...\left(general\right),$$

$$\theta = w_L/f ...\left(\text{for small divergence}\right).$$

A typical practical value that helps to describe the laser beam is the definition of the Rayleigh length z_R

$$z_R = \frac{\pi \cdot w_0}{\lambda}.$$

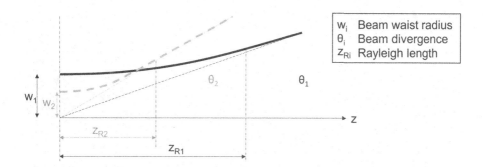

FIGURE 2.25   Beam caustic characteristics at the same BPP.

The Rayleigh length z_R is the distance along the propagation direction from the beam waist to where the beam area is doubled. The principal relations of beam caustic characteristics at same *BPP* are shown in Figure 2.25. At a constant *BPP* of the provided laser beam, a smaller beam radius will lead to a higher divergence angle and thereby to a smaller Rayleigh length.

The standard ISO 11146 requires 20 measurements of the beam radius along the optical axis (Figure 2.26). Ten measurements must be done around the beam waist within the Rayleigh length in order to detect the beam waist radius and its z-position precisely. The remaining ten measurements must be conducted in the far field in order to be able to detect the divergence properly. With these two parameters, the laser beam caustic can be described.

Laser beam measurements can be done with different methods. For rotational symmetric laser beams, non-contact measuring methods are possible, observing the laser beam from the side and recording the intensity projection. Other methods are based on measuring the spatial intensity distribution plane by plane, which enables the visualization of beam profiles. Rotating needles that raster through the laser beam and measure the local intensity along different paths through the laser beam

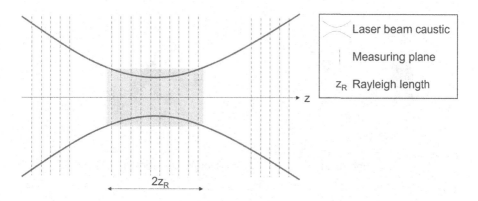

FIGURE 2.26   Sketch of a beam caustic, including minimum required measurement planes according to the standard ISO 11146.

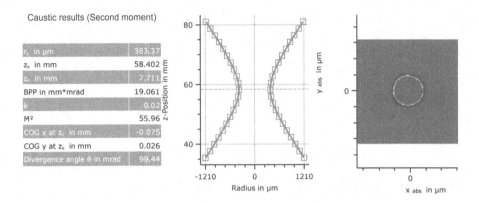

Caustic results (Second moment)	
r₀ in μm	383.37
z₀ in mm	58.402
zₐ in mm	7.711
BPP in mm*mrad	19.061
k	0.02
M²	55.96
COG x at z₀ in mm	-0.075
COG y at z₀ in mm	0.026
Divergence angle θ in mrad	99.44

FIGURE 2.27 Example measurement of the laser beam caustic and laser beam spatial intensity distribution at the beam waist (1200 W laser power, 1070 nm wavelength).

are combined to a distribution frame. Furthermore, camera chips can be used to record a spatial intensity distribution. The beam diameter is defined by the second moment method or the simplified definition that 86.5% of the laser intensity must be inside the laser beam spatial intensity profile.

All methods have in common to measure the beam radii/diameters in different positions along the propagation axis (Figure 2.27). The theoretical beam caustic equations are used to describe and visualize the beam caustic. In addition, laser beam characteristics, such as the z-position of the beam waist, the *BPP* as well as M^2, can be derived.

2.4   BEAM GUIDING AND BEAM SHAPING

The laser beam produced in the laser machine needs to be further delivered to the processing zone to enable material processing (Figure 2.28). Depending on the used

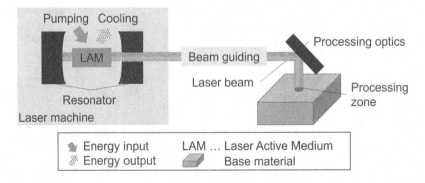

FIGURE 2.28 Sketch of the laser machine and principal path of the laser beam for its transfer to the processing zone.

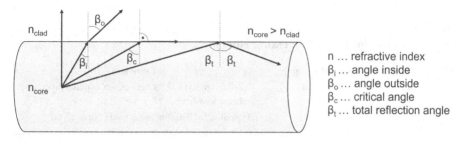

FIGURE 2.29 Visualization of light rays' interaction at the interface between two materials with different refractive indexes.

laser system, beam guiding can be done using mirrors or fiber optics. Further, optics are usually applied to project the laser beam onto the processing zone.

2.4.1 FIBER OPTICS FOR BEAM GUIDING

Laser light can be transported in optical fibers when the laser wavelength allows a high transmissivity through the fiber material. Wavelengths of the typical introduced solid-state laser emitting at ~1 μm wavelength show a high transmissivity. However, the CO_2 laser wavelength of 10.6 μm cannot be transferred in optical fibers.

Compared to free beam laser transport, fibers are more flexible and require less optical components. Little effort is necessary for the installation of machine systems, and a constant beam diameter is guaranteed at possible long fiber lengths. Laser light is guided in the fiber core by total reflection at the interface between the core with the optically denser material and the cladding (Figure 2.29).

The maximum angle β_t to achieve total reflection is defined by the refractive indexes of the used materials for the core and the cladding (clad)

$$n_{core} \cdot \sin(\beta_i) = n_{clad} \cdot \sin(\beta_o).$$

For enabling the coupling of the laser light into the fiber, the maximum angle α_{max} must be considered, defining the maximum Numerical Aperture *NA* (Figure 2.30)

$$NA = n_0 \cdot \sin(\alpha_{max}).$$

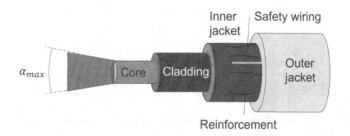

FIGURE 2.30 Sketch of the structure of a laser light fiber.

TABLE 2.6
Typical Values for Laser Processing Fibers

Characteristic	Typical Range	Comment
Core diameter	0.1–3 mm	0.2–0.6 mm (typical for high-power material processing, almost loss-free)
Fiber length	up to 30 m	Typically, 5–10 m (for single mode fibers <5 m)

In case the incident laser beam is larger than the *NA*, not all laser rays can be coupled into the fiber, and power losses occur along with heating of the fiber cladding. For a typically used fused silica fiber material, the *NA* is between 0.17 and 0.25.

Fiber bending partly decreases the incident angles of the laser rays inside the fiber core when interacting with the cladding and can be seen as a quasi-increase of the refractive index n_2, leading to a reduced maximum *NA*. In addition, polarization effects can occur affecting the output laser beam. Typical characteristics of fibers for laser beam guiding are summarized in Table 2.6.

Multi-mode fibers typically homogenize the spatial laser intensity distribution, leading to a typical top-hat beam profile (Figure 2.31a), while special fiber designs offer further possibilities of laser beam shaping in the fiber optic (Figure 2.31b and c).

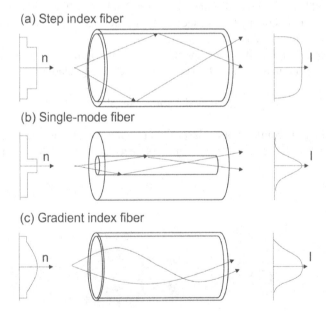

FIGURE 2.31 Different fiber types with different radial refractive index distributions *n* and intensity *I* outcomes.

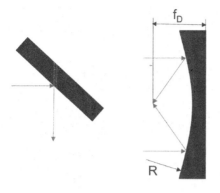

$$f_D = -\frac{R}{2}$$

R ... radius of curvature
f_D ... focal distance

FIGURE 2.32 Sketch of mirrors for laser beam shaping for direction change (left) and focusing (right).

2.4.2 BEAM SHAPING ELEMENTS

2.4.2.1 Mirrors

Mirrors are reflective optical elements that can be used for the change in direction of laser beam (Figure 2.32, left) or beam shaping, including focusing (Figure 2.32, right).

When selecting the mirror dimensions, the whole laser beam should illuminate the mirror area to avoid a cut-off of the laser beam along with a reduction in the transferred power and beam quality M^2. Mirror materials are typically chosen to show high reflectivity of the used laser light in combination with reflective coatings. Copper fulfills the requirement of high reflectivity for the IR wavelengths used for typical process lasers and offers in addition a good heat conductivity that enables quick heat transfer of the absorbed laser energy. Copper shows good processability that enables a high precision of the surfaces and a low surface roughness with possible cooling channels inside the mirror element. Cooling can be enabled by integrated cooling channels using cooling water, which avoids severe deformations of the reflecting surface even at high laser power illuminations.

Besides copper, silicon and molybdenum are used as substrates, which are also often used for beam deflection mirrors. Coatings are used to protect mirror surfaces against oxidation and enhance reflectivity to possibly >99.7%. A single layer anti-reflective coating with a layer thickness of $\lambda/4$ can be used (Figure 2.33a). For increased reflectivity, an accumulative high-reflection coating can be produced with alternating high- and low-refractive index coatings (Figure 2.33). Molybdenum mirrors can be used uncoated.

Typically, free beam guiding from the laser machine to the processing head is enabled by a combination of mirror systems. Such setups are mainly used for CO_2 laser light that cannot be transferred in fibers. Mirrors can be used for laser beam collimation and focusing.

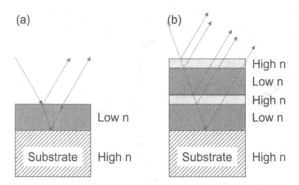

FIGURE 2.33   (a) Single layer anti-reflective coating and (b) multi-layer, high reflection coating based on different refractive indexes *n*.

2.4.2.2   Lenses

Lenses are transmissive elements that are typically used for guiding and shaping the laser beam (Figure 2.34). Plano-convex, convex-concave, and aspheric lens surface designs are used. Lenses are usually used at an angle of incidence of 0°.

Typical lens materials for different laser types are listed in Table 2.7. The surface quality of both surfaces must be in the range of $\lambda/20$. Typically, anti-reflective coatings are used for both interfaces in order to decrease the energy absorption of the optical element.

The central heating by the laser beam usually leads to a temperature increase and a thermal gradient in the lens material (thermal lensing effect). This can lead to a focus shift. Different materials show different thermal focus shifts (Figure 2.35) [Las21a]. The thermal impact of laser illumination on the optics needs to be considered when positioning the laser optics with the material to be processed to guarantee the correct laser beam diameter on the material surface.

Compared to mirror beam shaping, typically lenses offer the possibility to achieve smaller focusing spots and are less sensitive to deviations of the incident laser angle. However, lenses are more difficult to cool, which results in limited laser power that can be used. Anti-reflective coatings can be sensitive to mechanical impact, e.g., during cleaning.

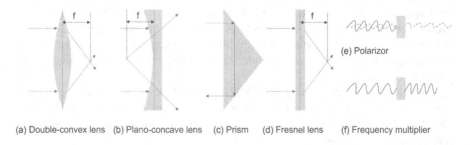

(a) Double-convex lens   (b) Plano-concave lens   (c) Prism   (d) Fresnel lens   (f) Frequency multiplier

FIGURE 2.34   Examples of different transmissive elements for beam shaping and guiding.

TABLE 2.7
Lens Materials for Laser Processing (Selection)

Lens Material	Comment
ZnSe	Suitable for, e.g., far-infrared-lasers
UVFS - Ultraviolet degree fused silica (SiO_2)	Suitable for, e.g., near-infrared-lasers
CaF_2 - Calcium fluoride	Suitable for, e.g., ultra-violet excimer lasers
BK7	Uniform transmission in visible range

Due to the cooling possibilities that are limited to the circumference of the optic, lenses often show thermal lensing effects when the temperature of the lens center reaches higher temperatures than that of the outer parts. This non-uniform (gradient) heating due to absorption of high-power laser radiation in optical elements causes thermal lensing, paraxial focus shift, and aberration. These effects lead to changes in the size and intensity profile of the focused spot in the optics. The induced thermal expansion in combination with the increase in the central temperature-dependent refractive index can lead to a decreased focal distance and increased laser spot size, which can affect processing. This effect leads to a change in the wavefront of the focused laser beam. In addition, as a consequence, a shift of the paraxial focus occurs, together with the appearance of the effect of spherical aberration, when different rays of the beam cross the optical axis at different points (Figure 2.36).

Fortunately, the physical properties of some natural optical materials allow either self-compensating or minimizing the optical effects induced by gradient heating, e.g., CaF_2 and crystalline quartz (Figure 2.35) [Las21a].

A remarkable property of crystalline quartz is that when the main axis of a crystal is orthogonal to the optical axis of the optical element (as in a waveplate), the self-compensation condition is almost exactly satisfied, and neither thermal focus shift nor aberration effects occur even in the case of contamination or surface absorption of optical elements. Users of optics made of crystalline quartz simply do not observe these undesired optical focus shift effects.

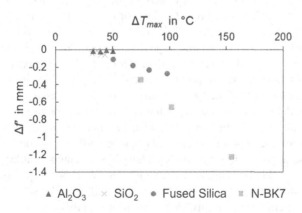

FIGURE 2.35 Thermal focus shift $\Delta f'$ depending on temperature increase ΔT_{max} [Las21a].

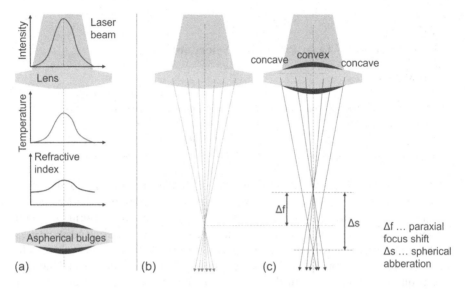

FIGURE 2.36 (a) Schematic sketch of the effect of the laser beam heating on the temperature distribution, the refractive index distribution and geometry of a lens; principal sketches of laser beam focusing of (b) a low-power laser beam without thermal lensing and (c) a high-power laser beam showing paraxial focus shift and spherical aberration.

Athermal crystalline quartz and sapphire with extremely high thermal conductivity ensure minimal temperature gradients. Optics made of these materials exhibit minimized thermal focus shift and aberration even during absorption of laser energy in the bulk material and coatings by contamination, scratches, and other surface defects. Weak birefringence of crystalline quartz and sapphire does not prevent their successive use in laser optics.

2.4.3 LASER BEAM FOCUSING

In order to utilize the laser beam for material processing, the characteristics of laser beams need to be understood. From the fiber or directly from the laser source, the laser beam shows a large divergence angle. Therefore, a collimator is typically used to guide the laser beam at a comparably large diameter and small divergence further to the focusing optic. Since the laser beam diameter on the focusing optic related to the focal distance of the optic is an important factor, the F-number is defined (Figure 2.37).

The principal impacts of the beam quality M^2 on the processing beam are shown in Figure 2.38. A better beam quality (smaller M^2) enables the use of smaller lenses at constant focal diameter, processing at a further distance at constant lens dimensions and focal diameter or a smaller focal diameter at the same focal distance.

When the same focal diameter and focal length is accepted, smaller optics (lower beam diameter on the focusing lens) can be used at better beam quality (lower M^2). Smaller processing optics can be used for easier handling and faster movement with possibilities of better accessibility. A higher beam quality enables a longer focal distance at constant spot size and lens diameter.

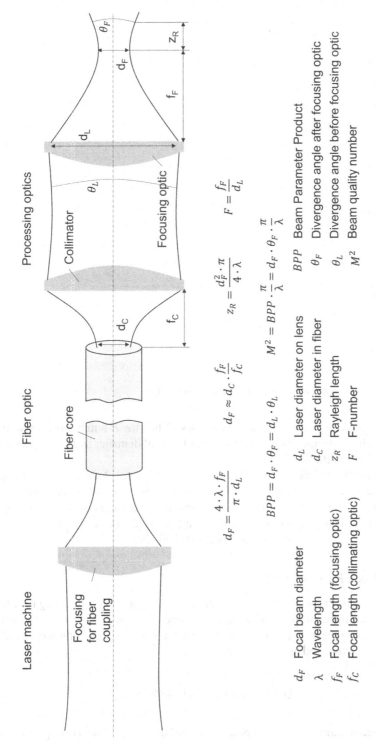

$$d_F = \frac{4 \cdot \lambda \cdot f_F}{\pi \cdot d_L}$$

$$d_F \approx d_C \cdot \frac{f_F}{f_C}$$

$$z_R = \frac{d_F^2 \cdot \pi}{4 \cdot \lambda}$$

$$F = \frac{f_F}{d_L}$$

$$BPP = d_F \cdot \theta_F = d_L \cdot \theta_L$$

$$M^2 = BPP \cdot \frac{\pi}{\lambda} = d_F \cdot \theta_F \cdot \frac{\pi}{\lambda}$$

d_F	Focal beam diameter	d_L	Laser diameter on lens
λ	Wavelength	d_C	Laser diameter in fiber
f_F	Focal length (focusing optic)	z_R	Rayleigh length
f_C	Focal length (collimating optic)	F	F-number
BPP	Beam Parameter Product		
θ_F	Divergence angle after focusing optic		
θ_L	Divergence angle before focusing optic		
M^2	Beam quality number		

FIGURE 2.37 Sketch of a typical path of laser beam guiding and forming, including main parameters and equations defining the laser beam.

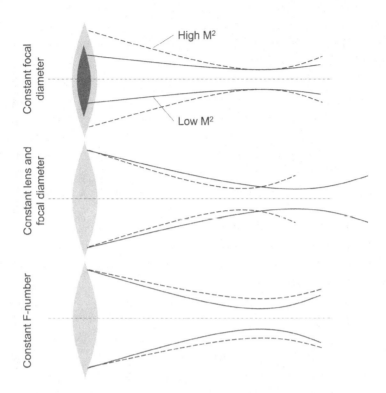

FIGURE 2.38 Impact of beam quality on the laser beam.

This feature enables remote laser processes. When the F-number is chosen to be constant, a better beam quality enables a smaller focal diameter and thereby a higher beam intensity in the focal spot.

The spatial beam intensity distribution typically changes along the beam propagation. In the beam waist, typically, the fiber end is projected. For typical multi-mode fibers, this is a homogeneous distribution (top-hat) over the beam area; for single-mode fibers, it is a Gaussian-like distribution (Figure 2.39). Further away from the beam waist, the intensity distribution becomes bell-shaped.

2.4.4 LASER BEAM SHAPING

For calculations, the perfect Gaussian beam is typically assumed, denoting the transversal electromagnetic mode (TEM_{00}). The intensity distribution $I(r,z)$ at every z-position along the laser beam axis of a round beam with radial symmetry in radius r direction is expressed as

$$I(r,z) = I_0 \cdot \left(w_0 / w_{(z)} \right)^2 \cdot exp\left(\frac{r^2}{w_{(z)}^2} \right),$$

with the beam radius in the focal position w_0 and the beam radius at the position $w_{(z)}$.

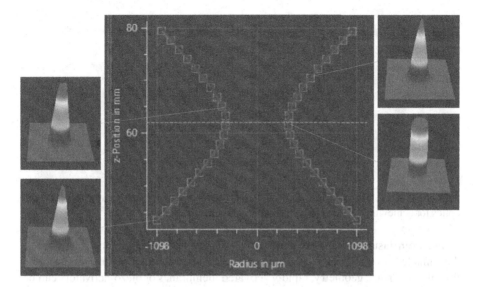

FIGURE 2.39 Example of spatial intensity distributions along the beam propagation of a multi-mode laser beam caustic (collimation lens: 150 mm focal distance; focusing lens: 250 mm focal distance).

Many high-power laser systems are multi-mode systems with $M^2 \gg 1$. Single-mode lasers are also available that typically achieve $M^2 > 1.1$. As already seen, the multi-mode spatial intensity distribution is typically top-hat-shaped in the focal position, where the fiber end is projected and changes into a bell-shaped profile when leaving the focal position. A single-mode laser beam keeps the bell shape at all axial positions. Beam shaping for high-power applications can be done using several methods that can be distinguished in dynamic and static beam shaping.

One possibility to dynamically shape the beam is the use of "flying" optics or also called scanning or oscillating optics. A typical setup is a combination of two moving mirrors that enable the laser beam positioning on the material. This method can be highly efficient for the so-called remote processing used for, e.g., welding, cutting, marking or powder bed fusion processing since the mirror movements are typically faster than computer numerical control (CNC) or robot machine movements that require positioning of the whole optics. The moving mirrors have much less weight to move to initiate the beam scanning. In practice, there are, of course, also limitations of the movement of the mirrors that, e.g., reduce the amplitude of the movement at high frequencies.

The common challenge of such systems is visualized in Figure 2.40a. The different path lengths s of the laser beam from the mirror to the interaction point with the substrate lead to different beam shapes and beam dimensions d on the substrate. Such a defocusing of the laser beam can lead to interruptions of the process and must be considered or avoided. Therefore, f-theta optics are used to guarantee that the laser beam focal position is, e.g., always placed on the substrate surface (Figure 2.40b and c).

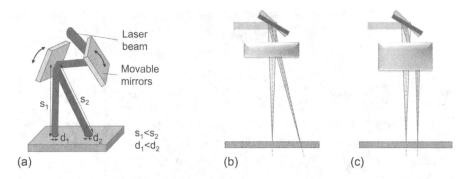

FIGURE 2.40 (a) Principal setup of a "flying" optics system and (b) and (c) with f-theta optics for achieving the same focal positions at beam deviations.

For even faster movements and spatial shaping possibilities of the laser beam, deformable mirrors are used. Adaptive mirrors using water/air pressure for changing the mirror surface geometry, rapidly activated membranes or piezo-activators can be used to induce rapid changes in the spatial laser intensity distributions.

The impact of an oscillating laser beam as heat sources for surface hardening is shown in Figure 2.41 [Dew20a]. In the turning points at the edges of the oscillation patterns, heat accumulation can happen that leads to increased hardened depth or local melting.

For static laser beam shaping, refractive or diffractive optical elements are typically used. Diffractive optical elements use micropatterns on the optic surfaces to create a variety of beam shapes in a flexible way using interference effects. Typically, thin optical material can be used, making the optics lightweight. Due to the fine surface structures, high laser power application was often critical to use to avoid melting. However, nowadays, possibilities increase.

Refractive optical elements use specially shaped surfaces of mirrors or lenses to guide the beam rays on the desired paths. Using such systems, e.g., single-mode

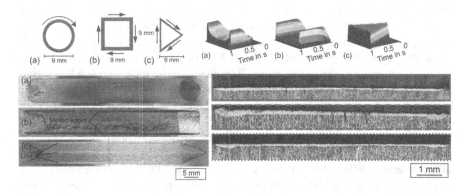

FIGURE 2.41 Surface hardening using beam oscillation for (a) circular, (b) rectangular, and (c) triangle oscillation (sketch of the patterns, temporal superposition, surface view, and cross-sectional view) [Dew20a].

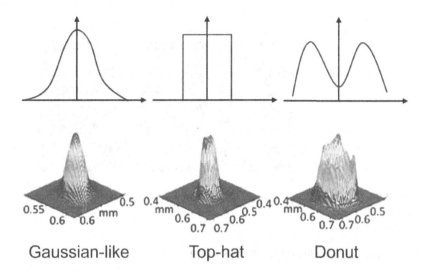

Gaussian-like   Top-hat   Donut

FIGURE 2.42 Refractive spatial beam shaping of a Gaussian-like beam into a top-hat and donut beam (top: schematic sketches; bottom: beam measurements).

beams can be formed from a Gaussian-like-shaped intensity distribution into a top-hat or donut beam shape (Figure 2.42).

Beam splitters enable the redistribution of intensity into several parts. Thereby, the beam is physically split in several profiles, e.g., to guide the beam around a wire feeder system. Another possibility of beam shaping is the beam splitting into several rings inside a fiber. Splitting the laser beam into several spots can also be accomplished with refractive optics (Figure 2.43). Multiple spot generation is possible with even controllable intensities for the single spots [Las21b].

In addition, several beams with different focal distances can be positioned along the beam axis, which enables to produce varied beam intensities along the beam axis and a "quasi-elongated" depth of field (Figure 2.44).

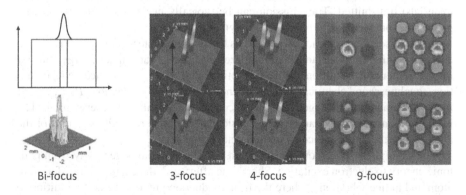

Bi-focus   3-focus   4-focus   9-focus

FIGURE 2.43 Beam shaping examples using superposition of multiple laser beams.

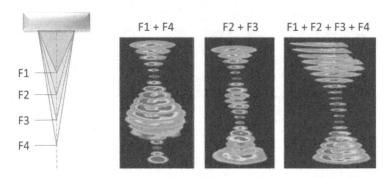

FIGURE 2.44 Axial superposition of multiple beams with focal positions *Fi* (adapted from [Vol21a]).

2.5 LASER BEAM ENERGY TRANSFER

2.5.1 PHENOMENOLOGICAL OVERVIEW

The laser is a thermal tool that basically only transfers energy to increase the temperature of material, unlike many other processing methods that use geometrical tools. The material impact depends on the interaction time and intensity of the laser beam, which describes the heat input. Absorption properties of the material are one of the main parameters that impact the energy input besides the surface geometry/quality. Laser light absorption depends on the material properties, inclination angle, surface temperature, and wavelength. The main heat loss of the processing zone is heat dissipation/conduction from the process zone, which depends on the material properties and part geometry.

Absorption happens when photons from the laser beam interact with the atoms (or molecules) of the material, while the photonic energy is assumed to be first transferred to the electrons that increase the energy state of the system. Afterward, the electron energy is transferred to the atom and the lattice, where heat conduction takes place in the form of particle collisions that transfer momentum and thereby energy. Laser energy that is not absorbed is either reflected P_{refl} or (in transparent materials) transmitted. Transmission can be typically neglected in metal material processing, but can play a role in, e.g., polymer processing.

While absorption is typically defined as the one-time interaction with a material, energy coupling or absorptance summarizes all effects that impact energy transfer into the material. In addition to sole absorption of a surface, geometrical (e.g., oxidation, roughness) and chemical aspects (chemical energy from reactions P_{chem}, e.g., of material elements with oxygen) need to be considered that alter the energy input. The most pronounced impact can be seen considering multiple reflections as part of the absorptance in a laser-induced vapor channel (keyhole).

The energy transport from a laser beam into the material is generally from photonic energy to electron excitation. Short time afterward, the energy is transferred to atom and lattice vibrations, where the heat conduction process starts. In addition to heat conduction, thermal radiation losses and convective losses can occur.

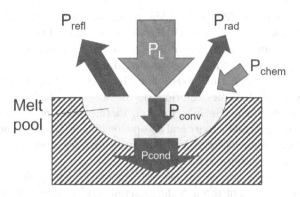

FIGURE 2.45 Energy transfer components in typical laser processes.

Thermal radiation (P_{rad}) can often be neglected in laser material processing due to its small portion compared to other energy transport phenomena. Thermal radiation follows the Stefan–Boltzmann law and is proportional to the fourth power of the material temperature. Heat conduction (P_{cond}) is the transfer of energy from atom to atom in a bulk material without changing the position of atoms. The process is temperature-dependent typically showing a slight decrease with increasing temperature, with severe changes at phase changes. Convective heat transfer (P_{conv}) plays a role in processes that include material melting and is defined as the energy transport due to change in atom position, and therefore, includes a mass transport also. In general, the liquid flow field depends on the temperature effects, geometry of the melt pool, and melt speed. An overview of the main energy transport effects is summarized in Figure 2.45.

The thermal efficiency η_{th} is defined as the power P_{th} required for melting the material volume in relation to the absorbed laser beam power P_A

$$\eta_{th} = \frac{P_{th}}{P_A} = \frac{A_{MP} \cdot \rho \cdot v \cdot \left(c_p \cdot (T_m - T_{amb}) + H_m\right)}{P_A},$$

with A_{MP} being the cross-sectional area of the melt pool, ρ the density, v the welding velocity, c_p the specific heat capacity, T_m the melting and T_{amb} the ambient temperatures, and H_m the latent heat of melting. The process efficiency $\eta_{process}$ is defined as the absorption efficiency η_{abs} times the thermal efficiency η_{th}.

$$\eta_{process} = \eta_{abs} \cdot \eta_{th}$$

The absorptivity A is the ratio of the power applied to the material to the power of the laser radiation

$$A = \frac{P_A}{P_L}$$

and can have values between 0 and 1.

The measurement of absorption has challenges due to many factors such as surface roughness, surface temperature, and wavelength and polarization of the incident light or oxide layers [Ind18, Küg19]. Absorption measurements have mostly been made on solid materials (e.g., [Dau93]). Lasers are often used as light source for the absorption measurements due to their distinct wavelengths, which makes it easier to relate an absorption value to a certain wavelength.

Absorption measurement methods mainly include calorimetric and radiometric measurements [Ber08]. Direct calorimetric measurements include thermocouples attached to the specimen [Het76]. However, thermocouple measurements require the knowledge of heat conductivity and dissipation effects [Gon07]. Therefore, water calorimetric measurements have been developed that are more accurate, but take some time to reach equilibrium states [Pep11].

Indirect radiometric measurements are suitable for absorption measurements when transmission of light through the material can be neglected, which is typically the case for metals in the dimensions relevant for high-power laser engineering. Kirchoff´s law correlates the spectral emissivity to the absorption. Integrating spheres (e.g., [Ber07]) are used, where the reflected laser light is absorbed and quantified. Furthermore, methods include gonio reflectometry, integrating mirror reflectometry, and emittance spectroscopy [Zhu04, Mod03, Tho82].

However, measurements at higher temperatures are difficult. Absorption values are often approximated at higher temperatures based on linear extrapolations.

2.5.2 Absorption of Metals

The absorptivity is a value that does not contain the absorption mechanism into the material. In case of metals, absorption takes place in a thin surface layer. In general, absorptivity depends on material parameters such as the refractive and absorption indices, electric and heat conductivity, and the specific heat. Knowing these parameters in combination with the laser wavelength and polarization, the absorptivity can be approximately calculated. Besides the material parameters, absorptivity depends on surface conditions (e.g., roughness), material geometry (e.g., material thickness), and applied or ambient gases. Furthermore, phase changes and vapor or plasma formation can influence the absorbed laser energy.

In general, metals are highly reflective, which means that laser light absorption is low. However, absorption varies with the wavelength of light (Figure 2.46). The

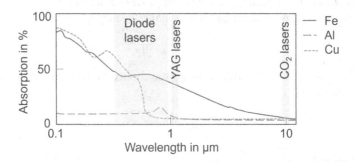

FIGURE 2.46 Wavelength-dependent absorptivity of different materials at room temperature and perpendicular laser beam illumination.

material dependence of the absorption of laser energy in different materials is shown in Figure 2.46 [Spi69]. Modern laser systems at 1 μm wavelength show a tendency to higher absorbance compared to CO_2 lasers (wavelength 10.6 μm) on typical materials for laser material processing.

Electromagnetic waves can be described using the Maxwell's equations. Beer's law explains the decrease of laser beam intensity in an absorbing medium along the propagation direction z to be exponential

$$I_{(z)} = I_0 \cdot e^{(-\alpha \cdot z)},$$

with the absorption coefficient α, intensity $I_{(z)}$, and the laser beam intensity at interaction I_0. The intensity of the laser beam is an important value for laser users defined as the laser power P_L related to the interaction area A_{beam}

$$I = \frac{P_L}{A_{beam}}.$$

Absorption length is the distance from the surface until the energy is absorbed. The absorption length l is defined with the absorption coefficient α as

$$l = \frac{1}{\alpha}.$$

The typical absorption length of laser beams into material is a few nanometers. Due to the short interaction distance, the laser beam can be typically well described as surface heat source. Since the interaction time is also short (~1 ps), the heat conduction is much slower and can be described with the classical heat conduction theories.

The absorption coefficient α is defined as

$$\alpha = \frac{4 \cdot \pi \cdot n \cdot k}{\lambda_0},$$

with n being the refractive index, k the absorption or extinction index, and λ_0 the vacuum wavelength. Table 2.8 shows some absorption and heat conductivity values for selected materials.

TABLE 2.8

Comparison of Absorption Coefficients α (at Different Wavelengths) and Heat Conduction Values

Metal	α (10 μm)	α (1 μm)	Heat conductivity (W/mK)
Copper	1.5 %	2 %	390
Aluminum	1.8 %	5 %	220
Iron	3.5 %	30 %	70

The mechanism of the absorption of laser energy in a surface layer is already under investigation for a long time and theories have continuously been improving. One of the first models was the Drude's model, which was developed based on the assumption that the free electrons are accelerated by the electric field and damped by collisions with phonons, other electrons, and lattice imperfections. Other theories, e.g. the Drude's theory, are based on solid-state absorptivity [Dru00]. The Drude's theory is based on classical physics and enables a good qualitative description of the absorption in many cases. One improvement of the model was the inclusion of the Fermi-statistics (e.g., [Pri95]).

The Fresnel equations explain the impact of the laser polarization, laser beam angle, and material properties on the absorption at a material interface with two different refractive indices. The absorptivity of an electromagnetic field (Maxwell's theory) is

$$A = 1 - R = 1 - \frac{(n-1)^2 + k^2}{(n+1)^2 + k^2},$$

with A being the absorptivity, n and k are the real and imaginary parts of the refractive index, and R the reflectivity.

The fundamental Fresnel equations derived (p...parallel polarization; s... perpendicular polarization) are valid for $n^2 + k^2 > 1$:

$$A_p = \frac{4 \cdot n \cdot cos(\varphi)}{(n^2 + k^2) \cdot cos^2(\varphi) + 2 \cdot n \cdot cos(\varphi) + 1}$$

$$A_s = \frac{4 \cdot n \cdot cos(\varphi)}{(n^2 + k^2) + 2 \cdot n \cdot cos(\varphi) + cos^2(\varphi)},$$

with φ being the angle of incidence against the work piece surface normal.

For a CO_2 and YAG laser beam, typical absorption curves depending on the incident angle are presented in Figure 2.47. The absorption of the s-polarized part of the laser light decreases at increased incident angles, while the p-polarized part shows an increase with a peak at the Brewster angle, where the absorption shows a maximum value and the p-polarized light is hardly reflected at $tan \varphi_{Brewster} = n$.

Unpolarized or randomly polarized laser beams are typically described by taking the numerical average between p- and s-polarized absorption. At lower wavelengths, the Brewster angle decreases. At higher temperatures, in general, a higher absorption can be expected.

Typically, an increased absorption is expected at increasing temperature with a jump at the phase transition from solid to liquid. The absorptivity increases with the temperature. A slight increase can be observed when increasing the temperature below the melting point and a sudden increase occurs when the temperature is close to the melting point. It is assumed that the material in a semisolid state shows surface roughening, which leads to a rougher surface and therefore a

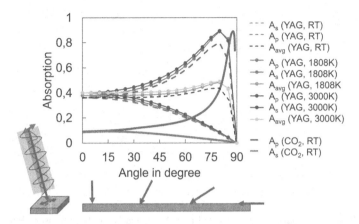

FIGURE 2.47 Linearly polarized (A_s...perpendicular; A_p...parallel; A_{avg}...average) YAG and CO_2 laser beam absorption, including room temperature (RT) and high temperature absorption calculations.

higher absorptivity due to multiple reflections on the rough surface (e.g. [Dau93]). Additional effects can alter the absorptance such as the creation of a plasma or scattering of light on small material particles in the vapor plume.

At higher temperatures, the laser beam absorption and further impacts of high temperature are widely unknown. The measurement of absorption on liquid or even boiling surfaces is challenging due to additional effects to consider such as thermal expansion, vapor ejection, and melt pool dynamics. Some attempts have been made using integrating spheres showing a slight increase in absorptivity at increasing temperature and wavelength [Bob80]. In addition, parallel temperature measurement must be conducted in order to relate the measured absorption to a certain temperature, which is challenging as well due to the often-missing knowledge about the actual emissivity of the material at high temperatures and the impact of high temperatures on the measuring equipment.

Recent radiographic measurements using a diode laser reflecting on a liquid metal surface confirmed the general trend of higher absorption at elevated temperatures [Vol23c] that interband and intraband absorption theories predict. Between melting and boiling temperatures, a local absorption minimum has been found, following the trend of combined intraband and interband absorption predictions (Figure 2.48). Above boiling temperature, the absorption values showed values closer to the sole intraband theory.

Around boiling temperature of steel, it has been shown that Brewster angles decrease to surprisingly low values that the Fresnel equations do not predict (Figure 2.49) [Vol23c]. Modeling the surface as a multi-layer structure with several interfaces between air, Knudsen layer, surface, and base material has helped to describe the observed trend. The results suggest that surface layering can possibly help to explain the absorptance effect at elevated temperatures.

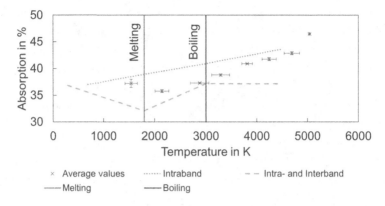

FIGURE 2.48 Comparison of measured and theoretically predicted absorption values at elevated temperatures (adapted from [Vol23d]).

2.5.3 HEAT TRANSFER IN MATERIALS

In laser material processing, the heat conduction equation is typically used to describe the thermal impact of the absorbed laser energy in a material. For solving the heat conduction equation, it is usually assumed that no convective or radiative losses appear since they are usually comparably low and that material properties are temperature independent. Common models can be distinguished in empirical, analytical, and numerical models. Empirical modeling uses input data from experiments to continue simulating certain aspects of a process. Analytical models use general relations between parameters to derive the results. More complex relations cannot be described with direct parameter relations but must be numerically solved in time-stepwise calculations. Such numerical models use typically finite element methods, e.g., for structural deformation calculations. Computational fluid dynamics models can also simulate fluid movement, which can help in describing the melt flow inside melt pools, spattering, or pore formation.

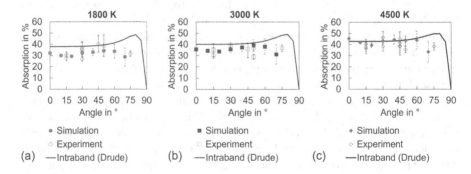

FIGURE 2.49 Absorptance measurements and simulation results at different incident angles and temperatures compared to the Drude's theory predictions (adapted from [Vol23c]).

Analytical solutions of the heat conduction equation exist for simplified bound-
ary conditions. When flat surfaces are irradiated, moving surface heat sources are
usually sufficient to describe the heat input into the material. The temperature field
resulting from a moving point source on a semi-infinite specimen can be described as

$$T(x,y,z,t) = \frac{1}{2\pi K \sqrt{x^2 + y^2 + z^2}} \cdot exp\left(-\frac{v}{2\kappa}\left(x + \sqrt{x^2 + y^2 + z^2}\right)\right),$$

with the process velocity v, thermal conductivity K, and thermal diffusivity κ
depending on the space coordinates and the time (x,y,z,t).

A line source on a semi-infinite specimen in x-moving direction induces the 2D
temperature profile

$$T(x,y) = \frac{Q}{2\pi K} exp\left(\frac{-vx}{2\kappa}\right) \cdot K_0\left(\frac{v\sqrt{x^2 + y^2}}{2\kappa}\right),$$

with the heat input Q in Watt per meter and the modified Bessel function of 0-order K_0.

A surface heat source on a semi-infinite specimen induces a 3D temperature profile moving in x-direction with the energy source in Watt per square meter and can be
evaluated numerically (e.g., [Mat10]). Thereby, the spatial intensity distribution can
be adapted. When a vapor channel is present in the process, other approaches can be
used to include the energy input in depth for numerical simulations, e.g., the Goldak
heat source [Kaz09] having a double ellipsoid heat input.

For finite element calculations, thermal field calculations enable the prediction
of thermal distortion, residual stresses, and the occurring phases. The used meshes
are typically larger compared to the fine meshes used for fluid dynamic simulations
and phase changes are often not directly modeled, which make the calculation time
comparably low. Fluid dynamic models require typically finer meshes and shorter
time steps in order to simulate material melting and even vaporization (e.g., [Ki02a],
[Ki02b], [Vol16]). However, even the creation of the keyhole and porosity, or spattering can be simulated [Ott11].

Latent heats of melting and vaporization are often required for process simulations but are typically difficult to integrate into the equations. The energy needed
during heating to overcome the bonding energy of the lattice is typically isothermal.
The same considerations are necessary during cooling.

Modeling approaches are partly highly advanced and can predict real-world relations quite accurately. However, often simulations are used to find working process
parameters or to increase knowledge about data that are not accessible or measurable.

2.6 MATERIAL PROPERTIES

Since laser material processing is based on the changes in material properties, a brief
section about some material characteristics during laser treatment is included here.
For more details, the reader is recommended to the respective specialized literature
given in this section. Typical materials for high-power laser processing are listed in
Table 2.9 as examples of a wide range of available, processable materials. In general,

TABLE 2.9
Selected Material Properties for Some Pure Materials

Metal	Density (g/cm³)	Heat Capacity (J/gK)	Heat Conductivity (W/mK)	Thermal Diffusivity (W/mK)	Melt Enthalpy (kJ/kg)	Vaporizing Enthalpy (kJ/kg)	Surface Tension γ_0 (mN/m) at RT	Surface Tension Coefficient $\Delta\gamma$ (mN/m)
Copper	8.96	0.39	394	1.14	205	4790	1330	0.23
Aluminum	2.69	0.9	221	0.91	397	10900	871	0.155
Iron	7.87	0.46	75	0.21	277	6340	1909	0.52

most of the material parameters are temperature-dependent. However, in special cases, the temperature-dependence can be neglected for modeling of processes.

2.6.1 SURFACE TENSION

An important material property regarding fluid behavior is the surface tension. It defines, e.g., the wetting characteristics [Eus13], the Marangoni flow, and melt pool dynamics [Sem06]. Surface tension is the resulting force needed to increase the surface area against the tangential bonding forces in a surface. It can also be defined as the energy needed to increase a surface area. Compared to a bulk atom/molecule that typically has twelve direct neighbors, there are only nine neighbors on the surface. This leads to the surface forces and the resulting surface tension [Mar11].

The surface tension is temperature-dependent. For pure materials, the surface tension decreases in general with the increasing temperature and can be, e.g., described as [Kee88]:

$$\gamma = \gamma_0 - \Delta\gamma \cdot (T - T_m),$$

with T_m being the melting temperature. This formula is mainly based on the measurements and calculations just above the melting temperature of the materials, e.g., maximum 2400 K for iron and steel (e.g., [Sey99, Mor11, Pok63]). However, this formula indicates that at a certain temperature, the surface tension disappears or even becomes negative. The main challenge of surface tension measurements are the conditions at such high temperatures that influence the measurements or affect the measurement equipment. Furthermore, liquid surfaces can become dynamic and show temperature gradients, which makes the definition of the actual temperature challenging. In addition, temperature measurement itself is a challenge at such high temperatures due to similar reasons.

Several surface tension measurement methods were developed (e.g., [Dre02]). For liquid metals, non-contact methods are preferred to avoid impacts on equipment. Therefore, oscillating drops can be used for measuring the surface tension, either during falling [Mat05] or more commonly while levitated [Tri88]. Also, sessile drop and pendant drop methods can be used to some extent. The maximum bubble pressure method enables measurements at high temperatures by evaluating the maximum pressure in a gas bubble pressed into the liquid material, but requires a tube physically moved into the liquid. Characteristic wave frequencies of surface waves traveling on the liquid metal can be related to the surface tension. Descriptions of the surface waves as capillary waves allow the calculation of surface tension depending on the wavelength and frequency (e.g. [Sch19]). Recent measurements using surface wave observations indicate that the decrease in surface tension is steeper than what linear approximations predict [Vol23e]. Above boiling temperature, the surface tension seems to stop decreasing (Figure 2.50).

2.6.2 PHASE CHANGES

Laser processing utilizes the thermal impact on materials. Materials can remain in solid state (e.g., during hardening), be transferred to liquid state (e.g., heat conduction

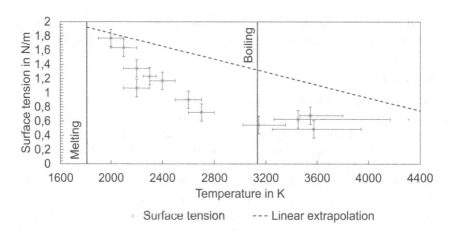

FIGURE 2.50 Surface tension of steel at elevated temperatures measured from propagating surface waves. (Adapted from [Vol23e].)

welding), or even be locally vaporized (e.g., deep penetration welding). In many processes, several phases appear at the same time, which means that thermal gradients occur; latent heats must be considered due to phase changes, and emissivity and absorptivity usually vary within the processing zone. Therefore, thermodynamic material properties play an important role in laser processing. Heat and mass transfer need to be known, e.g., for predicting porosity formation.

The mechanical properties and the material performance depend on the resulting microstructure that defines if the material is weldable. The microstructure mainly depends on the chemical composition and the thermal cycles applied. For the microstructural development, often the characteristic $t_{8/5}$ time (temporal duration during cooling of the material volume from 800°C to 500°C) is used to describe the thermal cycle and estimate the resulting microstructure from a thermal treatment (Figure 2.51).

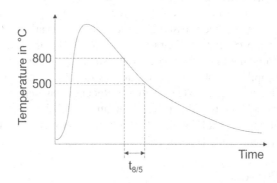

FIGURE 2.51 Typical thermal cycle during laser material processing with characteristic $t_{8/5}$ time.

In addition, laser processes induce material changes regarding microstructures. The heating and in particular the cooling conditions define the resulting microstructure after the thermal treatment. Typically, rapid cooling leads to the growth of comparably large grains. In steels, the transformation of austenite into martensite is induced. Phase changes can be desired (e.g., for surface hardening) or unwanted (e.g., when induced in the heat-affected zone during welding). Depending on the local thermal cycle that a volume element experiences, different microstructures can form. In general, larger grains are more brittle and can lead to cracking.

After a rapid thermal cycle produced by a laser beam, the created melt pool conditions often lead to dendritic grain growth during solidification.

In general, the resulting microstructure depends on

- the chemical composition of the material and its possible variations (e.g., element vaporization, chemical reactions),
- initial microstructure and precipitations,
- holding time above austenization temperature,
- heating and cooling rates, and
- melt convection and diffusion processes.

Different lattice structures of the same material are possible at different temperatures, which is the basis of surface treatments of metals. This process is typically accompanied by volume changes, induced stresses, and plastic deformation, which can in turn hinder the transformation or even cause cracks.

2.6.3 DIFFUSION

Element diffusion in a lattice mainly depends on the concentration gradient and the thermal cycle. The macroscopic diffusion can be described as

$$D = D_0 \cdot exp\left(-\frac{Q}{R_{gas} \cdot T}\right),$$

with T being the temperature, Q the activation energy, D_0 a factor, and R_{gas} the universal gas constant. Some typical diffusion parameters are listed in Table 2.10.

TABLE 2.10

Selected Diffusion Parameters of Several Elements in Different Base Materials

Material	Element	D_0 (cm²/s)	Q (kJ/mol)	T (°C)	D (cm²/s)	x (μm)
Fe	C	0.2	103	900	5e^{-6}	249
Fe	H	1.6e^{-3}	7.1	50	1.1e^{-4}	1170
Fe	V	3.9	242	600	1.3e^{-14}	0.01
Al	Cu	0.65	135	500	4.9e^{-10}	2.4
Al	Si	2.02	135	210	5e^{-15}	7.7e^{-3}

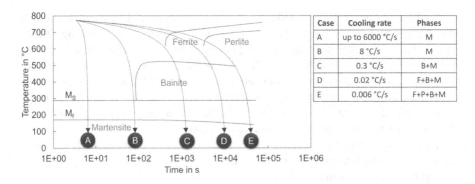

FIGURE 2.52 Schematic examples of continuous cooling of a C-steel at different cooling rates and resulting phases.

Diffusion depends on the materials and energetic situation describing how easy it is for an atom to leave its spot. The diffusion depth x_d can be estimated with the diffusion time t as

$$x_d = \sqrt{2Dt}.$$

Diffusion effects are, in particular, relevant for the short thermal cycle laser processes. E.g., in steels, carbon diffusion can be interrupted during rapid cooling, which can form phases that are not expected in equilibrium cooling conditions. At quasi-static conditions, where the steel has a very long time to cool, the Fe-C-diagram can be used to predict the resulting microstructure. However, when rapidly cooling, continuous-cooling-transformation (CCT) and time-temperature-transformation (TTT) diagrams (Figure 2.52) must be considered since non-equilibrium situations occur. Typically, the martensite start and end temperatures and austenization temperatures are increased at rapid thermal cycles.

2.7 PROCESS MONITORING

2.7.1 Process Emissions

Multiple emissions occur during laser processing that can be used to identify process characteristics. This information can be used for process monitoring and even for process control. Typical in-process emissions are shown in Figure 2.53, with typical ranges of wavelengths given in Table 2.11.

Optical emissions mainly originate from the reflections of the laser beam light on the surface, which can give indications about the process dynamics and thermal emissions due to heating. Acoustic emissions originate from the melt pool movement, vapor outflow and can transmit in the ambient gas or air or also within the material. Material is ejected in the form of spatters or vapor. Thermal radiation occurs from the heated surface.

Thermocouples can be installed to measure temperatures at fixed positions usually on solid material around the track area. Since the measurement happens further

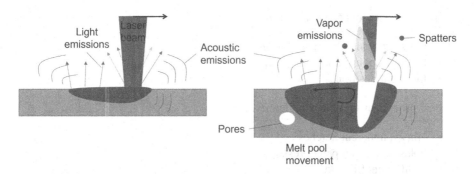

FIGURE 2.53 Schematic sketch of typical process emissions during laser beam processing without and with keyhole formation.

away from the melt pool, often thermal simulation models are calibrated to the measured temperatures to derive thermal information about the processing zone.

Radiographic thermal measurements do not require contact of the measuring device and are therefore useful to retain temperature data closer to the laser-induced track. Those sensors require translation of spectral emissions into temperature. Cameras or pyrometers use different wavelengths to analyze. One-color sensors need to be calibrated to give absolute temperature values but can already give useful information about temperature changes during a process. Multi-color systems evaluate minimum two wavelengths or wavelength range emissions. This can be done using a Dualscope [Dew20c], RGB-camera (e.g. [Vol23c]), or two-color pyrometers (e.g. [Gat14]). When the spectral sensitivity of the sensor is known, Planck´s curves can be used to calculate the temperature. In the calculation, black body radiation is often assumed, which might require calibration to an offset. The advantage of two-color systems is the independence from emissivity characteristics of the material.

2.7.2 PROCESS SENSORS

Process sensors can be categorized based on their application purpose. Pre-process detectors aim for detecting, e.g.,

- the position of joining partners,
- the gap and gap width between the joining partners, and
- the position of the tool center point.

TABLE 2.11

Typical Ranges of Process Light Emissions at YAG Laser Processing

	Thermal Radiation	Back Reflections	Plasma Radiation
Wavelength (nm)	900–2300	~1000	400–600

In-process sensors can detect, e.g.,

- the welding depth or root fusion,
- holes, dropouts, and pores, and
- the seam position.

Post-process evaluation can identify, e.g.,

- cracks, undercut notches,
- holes/pinholes (open pores),
- insufficient fill, and
- seam and edge quality.

Based on the emissions from the laser processes, sensors are chosen in order to get information about the process, while some sensors are used for pre-, in-, and post-process tasks (Figure 2.54).

Pre-process sensors are used to identify the correct positioning of the laser beam in relation to the specimen. This can be done, e.g., using visual cameras or triangulation sensors to identify the gap position for welding (Figure 2.55). During laser beam cutting, often induction sensors are used to maintain the distance between the sheet and the gas nozzle.

In-process monitoring is used extensively in research for detecting the spectral characteristics of vapor or plasma, temperature fields (thermocouples, thermal imaging), acoustic emissions for quantifying process dynamics or imaging system to detect geometrical and temporal events (Figure 2.56). Scientific methods include, e.g., high-speed imaging or inline X-ray CT. However, many methods are applicable to be also used in industrial environment and even usable for process control. Optical coherence tomography can, e.g., measure the keyhole depth during deep penetration

	Sensors		Purpose
Pre-process	Visual camera	Laser triangulation	Positioning Path planning Gap dimension
In-process	Visual camera	Spectrometer	Process imperfections Melt pool dimensions/dynamics Keyhole properties Material composition Thermal cycles
	Photodiodes	Acoustic sensors	
	Optical coherence tomography	Thermal sensors (camera, pyrometer, thermocouples)	
Post-process	Laser triangulation	Metallographic analyses	Material deposition Track dimensions/quality Imperfections
	Non-destructive testing		

FIGURE 2.54 Overview of typical sensors for laser process monitoring.

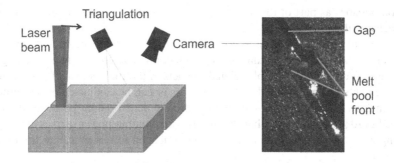

FIGURE 2.55   Sketch of laser triangulation and visual camera pre-process monitoring.

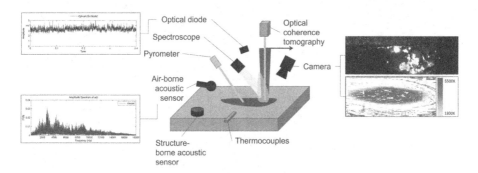

FIGURE 2.56   Overview of typical in-process sensors for laser processing and selected output examples.

welding, while deviating signals from the desired depth can be used to immediately change the laser power accordingly.

Post-process measurements are typical approaches to learn about the material and part characteristics after processing. An immediate measurement can be temperature sensors to record the cooling of the material or triangulation sensors to measure the geometry of the produced surface (Figure 2.57). Very typical are cross

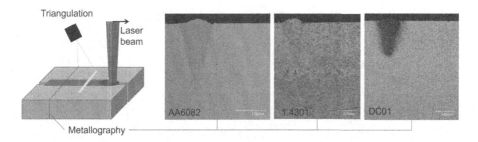

FIGURE 2.57   Post-process monitoring and metallographic analysis examples of aluminum alloy AA6082, stainless steel 1.4301, and DC01 steel.

section polishes as part of the destructive testing to learn about the microstructure or mechanical properties.

Additional sensors can be used to monitor equipment. Laser beam measurements should be conducted on a regular basis to detect possible contaminations along the laser beam path and the possibly related impacts of thermal lensing effects. Caustic measurements can be included into the process or can be conducted during process downtime. In addition, power meters measure the actual output power of the laser beam. In order to make sure that the programmed output power arrives at the processing zone, power measurements are recommended to integrate into the regular processing.

3 Laser Material Processes

The laser beam as a tool for material processing provides several advantages. Due to the local heat input, energy can be applied where needed and precise processing is possible. In addition, the high local energy input enables comparably fast processing. Laser processing is contactless, which means that the "laser tool" itself needs no maintenance. Modern beam guiding and optic designs make it possible to use the laser in automated ways, e.g., by robot handling. The high available beam qualities enable remote processing also, where optics can be positioned further away from the processing zone, which still reach the necessary intensities on the material's surface. However, laser systems can be a large investment and must be evaluated for each case if, e.g., the lower processing time offers the desired benefits compared to conventional methods.

Laser systems are already an important part of modern production. They show the ability to be integrated in automated production lines, and the flexibility to be used for many applications and processes. The laser is an unavoidable tool to enable the changes required for the implementation of Industry 4.0. The industrial laser applications 2017 were distributed to cutting (35%), welding/brazing (16%), marking (15%), semiconductors (14%), fine metal processing (8%), non-metal processing (6%), and additive manufacturing (AM) (4%). Descriptions of selected high-power laser processes is made in the following paragraphs.

3.1 OVERVIEW

Laser processes can be categorized based on the applied intensity and interaction time of the laser beam with materials (Figure 3.1). The chapter starts with the processes that operate at low intensity and take large interaction time; such processes include hardening, forming, and marking. At higher intensities, brazing and welding are possible, while at very high intensities, deep penetration welding, cutting, and drilling are enabled. In the end, the process variants of alloying, dispersing, and cladding are discussed, which lead to the AM technologies.

A laser beam carries energy in the form of photons. The lower the wavelength (higher frequency), the higher the energy per photon. When a laser beam interacts with a metal surface, energy is either absorbed or reflected (see Section 3.5.2). Therefore, laser beam is typically used as a thermal tool in materials processing to heat the surface of materials. The transferred energy to the materials and the induced heating defines the achieved processes.

At low intensities, only the surface of a material is heated (Figure 3.2). At higher intensities, the interaction may result in melting of material and even in vaporization. At very high intensities, plasma can form, which can be very hot but reduces the transferred laser energy to the material. Roughly as a thumb rule, intensity required

DOI: 10.1201/9781003486657-3

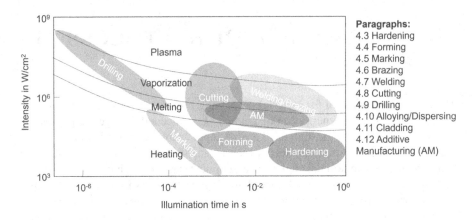

FIGURE 3.1   Rough categorization of selected high-power laser processes.

for metal heating is about ~10^3 W/cm² and vaporization starts at ~10^6 W/cm² for CO_2 lasers and at ~10^4 W/cm² for lasers with around 1 µm wavelength. As described earlier, absorption effects depend on many parameters, which makes an accurate prediction of the regime difficult, but can help to have a good starting point when developing a process.

Material properties (absorption, heat dissipation, etc.) and part geometries define the material reaction to laser heating. The main process parameters typically used for developing laser processes are

- intensity (power transferred per illuminated area) and
- line energy (energy input per traveled distance).

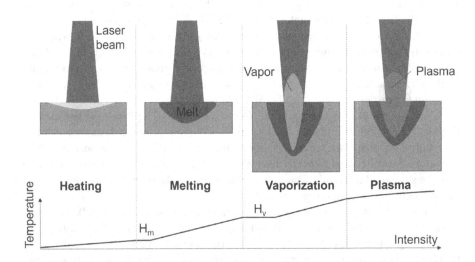

FIGURE 3.2   Laser beam heating regimes depending on the laser beam intensity.

Typical advantages of laser processing technologies are high precision and comparably low, but local energy input, while no direct mechanical forces are necessary. This can lead to short production times and high quality. Processes can be highly flexible with respect to geometries and materials. Process efficiency can be improved compared to other processes, although high wall plug energy is typically required to run a laser. When the processing speed is high enough and quality requirements cannot be fulfilled otherwise, the laser is the ultimate tool to use.

Laser processing heads are usually mounted on a robot or a computational numerical control (CNC) machine due to the required precision of positioning and high processing speeds. Nowadays, industrial applications are installed in mass production for repetitive tasks. However, the possible advantages of tailoring the laser beam to customized processes are in progress.

3.2 SAFETY

When working with lasers, precautions are necessary. Lasers are classified into four laser classes. Only class 1 lasers are not harmful to humans. Class 2 lasers are in the range of 0–1 mW, followed by class 3R (1–5 mW) and 3B (5–500 mW) lasers, where an eye injury risk is medium/high. Severe damage must be expected when using class 4 lasers. Lasers above 0.5 W in continuous wave (cw) as used in high-power laser processing are all class 4. For increasing safety, processing cells can be installed to make the processing area sealed from the outside. When using the correct housing, the operators outside are safe from laser impacts, denoting a class 1 zone outside.

All laser machines must be labeled with laser class. Laser warning signs need to be installed to make people aware of the risk including laser wavelength and maximum output power. For protection of humans, safety for eyes and other precautions need to be provided according to the risks. Not only directed laser light can be dangerous for the human eye but also reflections (diffuse or on reflective surfaces) can be dangerous. Even a collimated laser beam focused on a small area inside the eye can increases its intensity and cause irreparable damage. This also happens in case of invisible infrared and near-UV light. CO_2 laser light (mid-infrared) is mainly absorbed in the outer cornea and might burn those parts.

Therefore, wearing appropriate laser safety goggles for all emitted wavelengths is mandatory (indicated by a high-optical density, OD, number for the emitted wavelength), which usually protects the eye from a 10 s collimated beam, but there is still a high risk when illuminated by a focused beam. The goggles should have no scratches or damage.

Further protective measures should be installed such as passive walls that show slow melting at irradiation. This gives operators a chance to press an emergency stop button. Active walls can initiate an emergency stop due to their integrated sensors for automatic shut-down at too high energy illumination. Warning lights, key switches, warning signs, and safety training complement a safe laser handling.

Secondary risks like burning of material or creation of fumes and toxic gases must be considered when creating the setup, and suitable protection mechanisms need to be in place, e.g., exhaust systems, shielding, and protection gases.

3.3 SURFACE HARDENING

3.3.1 OVERVIEW

Laser surface hardening is a process within the category of surface treatment, where the material is heated to temperatures typically below the melting point to initiate phase transformations that can alter the mechanical properties of the surface volume. No additional material is needed for this process. The laser beam is used as the heat source to initiate the surface heating. Since usually no melting is involved, shielding gas to protect the processing zone from the environment is not always mandatory. For some materials, oxidation effects need to be avoided by applying shielding gas.

Typically, low laser beam intensities (~10^3 W/cm²) are sufficient to induce surface hardening. For this, CO_2 lasers, solid-state lasers, and in particular, diode lasers can be used. The typical surface layer thickness that can be hardened is usually in the range of 0.5–1.5 mm. The main process parameters are laser power (typically in the range of 1–5 kW) and processing speed (typically in the range of 0.5–10 m/min), while at higher laser power and lower speed, there is a general tendency to wider and deeper hardened surface layers. Typical hardened surface regions are shown in Figure 3.3.

Since the laser beam is absorbed by a solid flat surface (Figure 3.4), the absorptivity is comparably low. The laser beam can be inclined in order to increase the energy absorption due to the general trend of increased absorption at higher incident angles. When using a polarized laser beam, this effect can be even enhanced (see Section 3.5.2). Additionally, surface preparation can increase absorptance, e.g., additional coatings or sand blasting. In general, rougher surfaces increase the possibility of laser rays to be absorbed and reflected from the surface to another area on the surface. Such multiple reflections and absorptions increase the energy input.

The shape of the hardened zone is mainly defined by the heat conduction of the absorbed energy on the material's surface. Therefore, the hardened zone typically shows a round shape following the isotherms created in the material. The appearance of hardened zones can be altered by changing the energy input or varying the intensity (laser power and process speed). Another way of influencing the shape of the hardened zone is beam shaping (see Section 3.4.4). Beam shaping in hardening is a

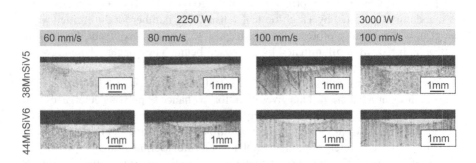

FIGURE 3.3 Cross sections of hardened surface regions at different laser powers and processing speeds.

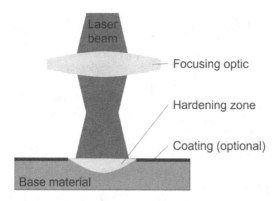

FIGURE 3.4   Principle sketch of a laser hardening setup.

common way to increase the hardened area, when wider hardened zones are desired. Since low intensities are needed, a simple defocusing widens the laser spot on the material (typical used spot diameters are in the range of 1–10 mm) and increases the hardened zone dimensions if the intensity or line energy can be increased enough to still induce the hardening process. In addition, beam shaping using diffractive or refractive optics or oscillation mirrors can tailor the local energy input. High-intensity peaks should be avoided in static beam shapes to avoid local melting. In dynamic beam shaping using oscillation optics, path planning often needs optimization. During the acceleration and deceleration of beam oscillation pattern, the laser beam illuminates sections for varying time durations. Therefore, in particular, at the turning points, melting can occur.

The laser beam enables a local energy input. Comparable effects in the material are induced during conventional hardening methods (e.g., induction, flame, or oven heating), but faster processes are possible using the laser beam technology. The local energy input leads to lower distortion effects compared to conventional methods, where the whole part or large sections are heat treated. The disadvantage can be comparably low hardening depths. Applications of hardening tasks can be found in tools, inside of cylinders, or for drilling holes.

3.3.2   MARTENSITE HARDENING

The main principle of hardening a surface is to induce phase changes in the surface-near area. In steels, martensite (transformation) hardening is typically applied. Carbon-steels with carbon contents between 0.4% and 1.5% work well to induce this effect. Thereby, the austenization temperature is aimed to be reached during the process in combination with self-quenching or induced quenching to form martensite (Figure 3.5) [Sch10]. The carbon in the ferrite structure leads to distorted microstructures that show high hardness. Compared to conventional hardening processes, the laser energy input can be tailored to reach temperatures just below the melting point.

The locally reached temperature and holding time define if phase transformation happens from face-centered-cubic (ferrite) to body-centered-cubic (austenite)

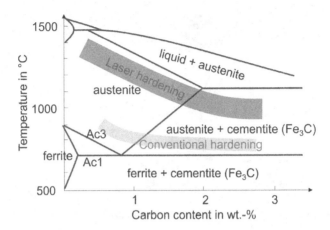

FIGURE 3.5 Principle sketch of a section of the Fe-C-diagram at equilibrium cooling with typical maximum temperatures of conventional and laser treatments.

and during cooling to the body-centered tetragonal structure (martensite) without significant carbon diffusion. Martensite has a higher volume compared to ferrite and austenite, which leads to residual compressive stresses during its structure formation. In addition to the martensitic hardness, those stresses can increase the hardness even more.

3.3.3 LASER MARTENSITE HARDENING

Hardening processes are in general well known and examined. However, the short thermal cycles during laser processing can lead to different behaviors compared to the regularly used slow cooling hardening processes. At the typical short thermal cycles when using laser processing [Gar19], fast cooling can induce martensitic phase transition and strengthening in carbon steels. Fast heating occurs in the range of ~100 K/s [Wis85], which also induces an increase in the transformation temperature (Ac1 and Ac3, start and finish of austenization). Time-temperature-transformation (TTT) diagrams help to describe changes due to faster heating compared to equilibrium heating (Figure 3.6). In laser beam processes, the cooling rates are typically so high at and around the processing zone that martensite hardening is induced.

Martensite hardening is based on the availability of carbon, which needs to be homogeneously distributed in the surface region to achieve a surface layer of homogeneous properties. However, the homogenization of carbon in the lattice takes time. Such a diffusion process is related to the carbon gradient in the material and exponentially to the temperature (see Section 3.6.3). Due to the induced short thermal cycles, even after reaching temperatures above Ac3, carbon may not be homogeneously distributed, which can lead to inhomogeneous microstructures and potentially lower hardness values than expected, since martensite transformation cannot be induced everywhere (Figure 3.7). In such cases, depending on the base material, austenite or retained ferrite grains can remain in the material. On the one hand, this effect can be considered unwanted since the intended homogeneously hardened surface is not achieved, and on the other hand, the creation of mixed structures opens

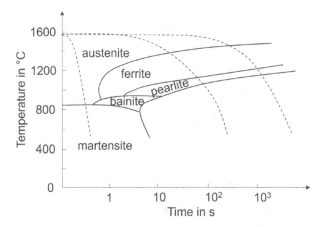

FIGURE 3.6 Principle sketch of a TTT diagram with different exemplary cooling curves.

new possibilities of tailoring the hardness and potentially other properties directly by applying the needed temperature field using the related laser parameters.

Fast cooling is typically induced by self-quenching. The hardened surface areas are typically small compared to the part dimension. Therefore, heat is efficiently and rapidly transferred away from the processing zone. Cooling rates of ~10 000 K/s are possible during laser processes. In order to further increase the cooling rate, external quenching can be applied. Using oil or water to rapidly decrease the temperature of the whole part or locally of certain volumes can support the hardening process. The complete thermal cycle defines the resulting hardness in every volume element (Figure 3.8). This includes the heating time, the holding time at high temperatures and the cooling rate.

A main difference of laser hardening compared to other hardening processes is that not the whole part or large sections are heat treated with the same temperature

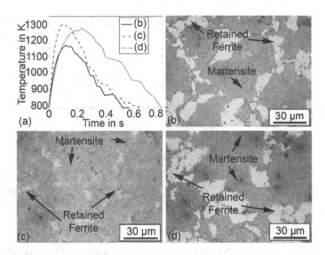

FIGURE 3.7 (a) Calculated thermal cycles in the respective areas: Microstructure after laser surface treatment at the (b) transition, (c) middle, and (d) overlapping areas [Dew22].

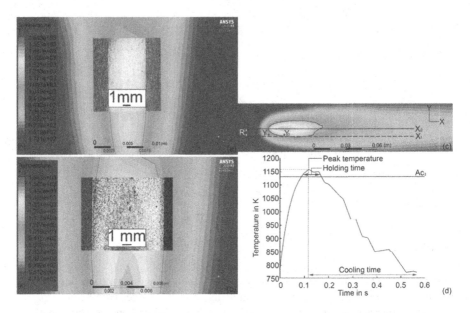

FIGURE 3.8 Examples of thermal simulations in top view including the related hardened tracks (a, b) and thermal cycles with heating, holding, and cooling times (c, d) [Dew22].

profile. The temperature history of each volume element of the heat-treated material can significantly vary in depth and distance from the heat source center. Therefore, it is necessary to consider that each volume element has a different thermal experience. Depending on the thermal cycle at the local volume, different microstructures can form at different locations (Figure 3.9). This can lead to complex hardness profiles in depth.

3.3.4 PRECIPITATION HARDENING

Besides martensitic hardening, precipitation hardening is an additional effect that can increase surface hardness (e.g., [Har95]). Thereby, addition of alloying elements,

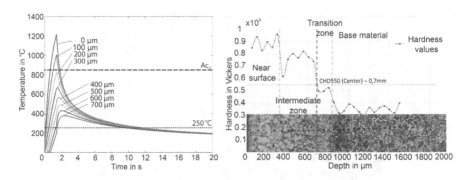

FIGURE 3.9 Thermal cycles (left) leading to different microstructural developments and hardness values in depth.

e.g., titanium, vanadium, and niobium, to the chemical composition can create fine precipitates (nitrides, carbo-nitrides) on mainly grain boundaries and suppress grain growth, which induces a higher hardness after cooling.

3.3.5 BEAM SHAPING FOR SURFACE TREATMENT

Beam shaping methods can be used for homogenization of the temperature field on the material's surface to use, e.g., higher laser powers without melting local surface and still reaching high enough temperatures on a large area. Typical beam shaping equipment include kaleidoscopes, mirrors, diffractive optical elements (DOE) including axicons or transmissive lenses for refractive beam shaping. Furthermore, beam oscillation optics are flexible tools to change the intensity profile with temporal variation, while a comparably small spot is moved over a large area at high speed by moving mirrors. Thereby, the oscillation frequency and amplitude define the spot movement and the heat input. The laser beam can be moved fast enough not to melt the surface, while the accumulated heat input is high enough to reach surface heating. In practice, mechanical limitations of the mirror movement need to be considered, since programmed patterns might not always be reached due to too high necessary accelerations. Beam shaping mainly changes the geometry of the hardened zone. A Gaussian-like, a square top-hat, and a 4 × 4 DOE spot distribution are compared in Figure 3.10.

Beam shaping technologies offer the possibility to produce different widths of the hardened zones and different temperature fields on the material's surface (Figure 3.11). The strong variations in the hardness values indicate that islands of retained ferrite with lower hardness values are present, in particular, in the transition zones from the hardened zone to the base material.

3.3.6 PRACTICAL ISSUES

While hardening curved surfaces, different absorption values must be expected in different areas due to different incident laser beam angles relative to the surface. Reflections from part surfaces can lead to unexpected local heat inputs. In particular, multiple reflections and additional absorption spots are possible that can locally

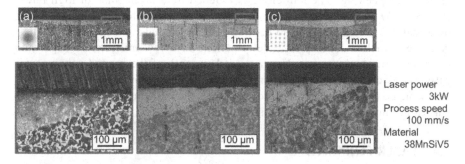

FIGURE 3.10 Hardened surface areas in cross-sectional view at (a) Gaussian beam, (b) square top-hat beam, and (c) 4 × 4 spot beam illumination [Dew20b].

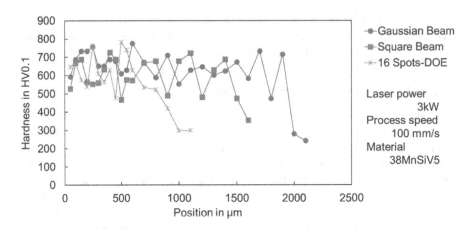

FIGURE 3.11   Hardness profile from the center of the hardened zone in horizontal direction.

increase the heat input, which in turn can lead to local melting. Such heat accumulation can happen, e.g., at crankshaft hardening, where the laser beam has to be transferred into the fillet with a steep angle. Beam shaping in combination with temperature monitoring and possibly control can counteract the problem of local overheating.

Another typical effect in laser hardening occurs while hardening large areas that require several hardened tracks next to each other. The previously hardened surface area is affected by the heat input of the subsequent tracks. Every subsequent heat input induces thermal cycles in the already hardened area that might not lead to the desired martensitic hardening again. Therefore, a soft zone can be created that locally reduces the hardness and can be a weak section where, e.g., fatigue cracking can be initiated. A similar effect occurs when one track is applied with the same start and end point, where the already hardened starting area is heat treated again by the process end.

3.3.7 LASER SHOCK PEENING (FOR HARDENING)

A special way of inducing deformation of a material to increase its hardness is laser shock peening. The hardening is induced by a laser pulse of high intensity (>1 GW/cm²), which creates a shock wave on the material's surface. Air or other gases are ionized. The so-created plasma expands and induces a surface pressure, which induces plastic deformations in a few millimeters depth and the related residual stresses. Hardness and strength of the treated volume typically increase and often corrosion resistance can be improved along with fatigue behavior. Laser shock peening processes are used for many purposes such as hardening turbine blades or plane structures.

3.4 LASER FORMING

Laser forming is a highly flexible thermo-mechanical process. The laser energy input is used to induce thermal distortion to shape the part. No additional hard tooling or external forces are needed to be applied. Solely the locally induced thermal field in the sheet produces the necessary deformation.

The laser beam caustic shows a small radial dimension, and the processing plane can be positioned far away from the laser optics. Due to these features, the laser beam can even reach part locations that are not accessible with conventional bending tools. The good accessibility of laser beams enables the complex shaping of many parts and components. All materials that can absorb the used laser beam can, in principle, be processed for laser forming. This enables a wide range of usable materials for forming procedures, e.g., the bending of otherwise hard-to-form materials.

The laser beam is typically defocused on the material's surface using low intensities that do not melt the surface. The creation of mushy zones or even melting destroys the bonds in the lattice and cannot lead to the desired controlled plastic forming. In a solid material, the induced thermal gradients into the material create the necessary stresses to lead to the plastic deformation needed to form sheets. The introduced thermal stresses in the material's surface induce plastic strains that bend the material along this path. Typically, the laser beam is moved along the bending line over the part to be shaped.

Depending on the induced temperature field in the material, different bending effects occur. The temperature gradient mechanism is initiated when mainly the upper part of the structure is heated (Figure 3.12a). This effect occurs when the sheets are thick compared to the impacted surface by the laser beam. The thermal expansion first leads to elastic deformation (counter bending) and partly plastic deformation. After turning off the laser beam, heat conduction induces contraction heating of the lower structure part and plastic deformation. During the remaining cooling time, the structure bends slightly back, releasing the elastic thermal deformation. Some typical surface occurrences and bended sheets are shown in Figure 3.13.

The buckling mechanism is induced when the laser beam leads to an expansion of the heated area and thermally induced contraction and the related material reaction (Figure 3.12b). In case of an inhomogeneous heating where heat is accumulated in the lower parts of the sheet, a bending of the sheet in the opposite direction can be achieved.

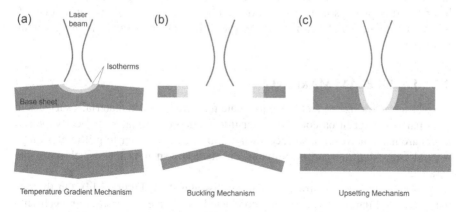

FIGURE 3.12   Forming mechanisms using a laser beam.

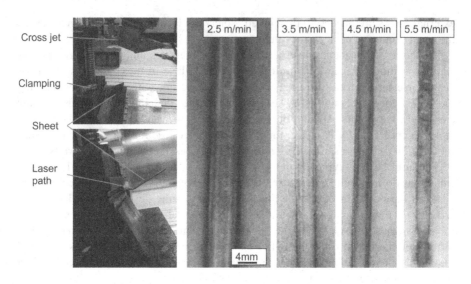

Cross jet

Clamping

Sheet

Laser
path

2.5 m/min   3.5 m/min   4.5 m/min   5.5 m/min

4mm

FIGURE 3.13 Processing setup including bended sheet and surface appearances at different single treatments at different processing speeds.

When the induced heat impacts the whole thickness of the sheet nearly homogeneously in depth, mainly an elongation of the sheet can be achieved. The plastic deformation is initiated along the sheet and after cooling and release of the elastic deformation, the sheet is elongated (Figure 3.12c).

Due to the complex stresses that may already exist in the sheets from the sheet production process, the result after laser bending is difficult to predict. In addition, complex stress fields can be induced due to variations in heat input or accumulations of heat at the beginning or end of the sheets or at cut-out parts in the sheet. Corrections after laser bending can be necessary. Based on the actual geometric measurements of the sheet after the process, counteracting laser treatments can be calculated and initiated in the next forming steps [Tho20]. The laser forming process is easy to control and can show good energy efficiency. It is, in particular, suitable for rapid prototyping and low-volume manufacturing processes.

3.5 LASER BEAM MARKING

Laser beam marking is a technique to create fine structures on material's surfaces to label parts or imprint barcodes or other information on products. Typically, pulsed lasers are used; however, a surface variation in the micrometer to millimeter range is produced. Marking can be initiated on materials such as metals, plastics, and ceramics.

Two main processing strategies can be distinguished (Figure 3.14). On the one hand, the masking method is usually conducted using an excimer laser and is highly

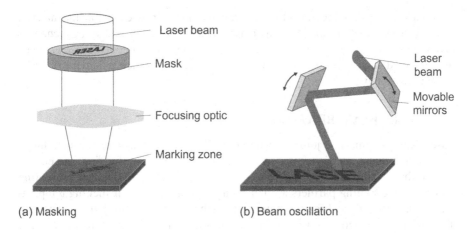

(a) Masking   (b) Beam oscillation

FIGURE 3.14 Laser beam marking processing strategies using (a) a mask or (b) beam oscillation.

efficient in mass production for printing the same information multiple times. On the other hand, beam oscillation is highly flexible and enables printing of customized information. The possibility of producing multiple markings next to each other enables larger area treatments.

When using pulsed laser beams, the pulse overlap needs to be considered as an important parameter that defines the resulting surface quality. Rippled structures might occur, which requires post-treatment to enable high-quality surfaces.

Three principal material reactions can be initiated by laser marking: (1) Removal of material creates grooves that can be distinguished from the rest of the material's surface, due to different reflections of the surrounding light; (2) color changes can be initiated either by thermal or oxidation effects on metals or by pigmenting in plastics; and (3) volume changes by forming voids can be induced mainly in plastic materials (Figure 3.15).

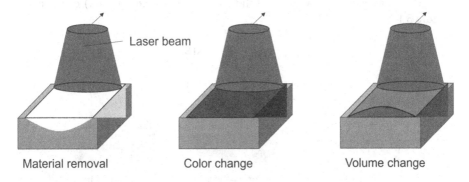

Material removal   Color change   Volume change

FIGURE 3.15 Material marking effects to create a marking track.

Removal of material requires typically high laser powers to evaporate the material and/or eject the melt as spatters. The main advantage of color and volume changes is that no grooves exist after processing. This can be a desirable feature, e.g., when marking medical devices that need to be properly cleaned, which is harder to achieve when grooves exist.

3.6   LASER BEAM BRAZING

Laser beam brazing is a joining technology. The joining partners are metallurgically joined using additional brazing material. In laser beam brazing, the energy to melt the brazing material is provided by the laser beam. In contrast to welding processes, the joining partners are not brought to or above their melting temperatures. Only the added brazing material is melted and wets the joining partners. Thereby, the connection of the joining partners is achieved. Therefore, the melting point of the joining partners should be higher than the one of the added material. Wire is typically used as feedstock for brazing. The wire is fed by the wire feeder to the processing zone. The laser beam provides the energy to melt the brazing material and also pre-heat the base material. Pre-heating can improve the wetting behavior and thereby increase the connecting area between the joining partners and the brazed material.

Shielding gas is typically used to avoid oxidation of the material. The heat capacity of the added material should be high and the heat conductivity of the joining materials should be low to enable an efficient process. These characteristics will ensure not to lose energy to the surrounding volume.

The bonding between the joining partner and the brazing material is achieved by diffusion and the initiated place changes of molecules and atoms, which is called cohesion bonding. It is typically not adhesion bonding when molecular bonding is initiated. For achieving a good bonding, the wettability and thereby the wetting angle θ play an important role (Figure 3.16). Material combinations with good wettability show small wetting angles and those with poor wettability show large wetting angles.

The Young equation describes the relation between the wetting angle θ_b to the surface energies γ between the solid (S), liquid (L), and gas (G) components (Figure 3.14)

$$\gamma_{SG} = \gamma_{SL} + \gamma_{LG} \cdot \cos(\theta_b).$$

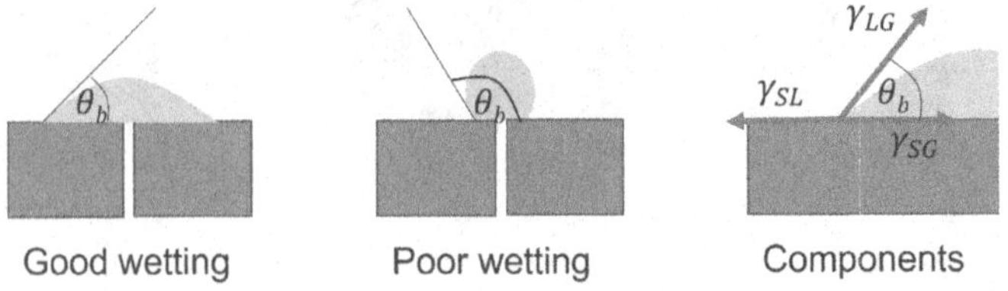

Good wetting   Poor wetting   Components

FIGURE 3.16   Wetting angles θ_b during brazing and surface energy components γ.

The surface tension of the liquid material is often known from literature or can be measured. The surface energy of the solid material of the joining partner can be typically also found in literature (e.g., [Kee88]). However, in order to derive the surface energy between the liquid and the surrounding gas, a wetting experiment is necessary. Typically, the wetting angle is measured in tests letting material drop and wet on sample base material. With the known surface energy of the solid and the surface tension of the liquid, the interface tension can be calculated.

Wetting requires a certain time until a stable shape is created. Since in laser processing, typically the time until solidification is relatively short, full wetting may not be achieved. Therefore, the calculated wetting angles might differ from the actual brazing wetting angles. Due to the induced temperature fields in the joining partners' material, wetting can be reduced or even stopped, when the base material is not warm enough for the liquid to continue spreading.

Wetting can be and often must be supported by applying flux in order to locally remove the oxide layer from the surface of the solid material. Fluxes typically consist of non-metallic material, e.g., silicates, carbonates, chlorides, which contain acids or build acids when heated that enable the mechanical removal or chemical reactions. Such fluxes have a low melting point (~50°C). Therefore, the temperature and time intervals of fluxes are limited. Alternatively, plasma cleaning supported by the kinetic energy of ions that break the oxide layer is possible. Thereby, the plasma process can be installed right before the brazing process in order to minimize the available time for re-oxidation. The integration of the plasma process is an additional investment and limits the processing flexibility due to intrinsic direction-dependence.

Typical applications of laser beam brazing can be found where an appealing appearance and moderate strength of the joint is necessary. Disadvantages of brazing compared to welding are

- the comparably low tensile strength due to the bonding type, since the joint is based on atomic movement and diffusion instead of mixing of molten materials,
- the possible cracking due to inhomogeneous cooling of materials and the creation of thermal stresses that do not release during cooling and lead to cracking,
- the tendency to higher corrosion probability due to the joining of dissimilar materials and the possibility of creating intermetallic phases that are prone to corrosion. In addition, microcracks from inhomogeneous cooling denote access of corrosion supporting liquids,
- the need for flux that denotes an additional process step for its application on the surface, while fluxes are often unhealthy for humans and special care needs to be taken during the mixing of components and also during the processing,
- the often-required preparation of surfaces before joining to enable sufficient wetting and bonding, and
- the formation of brittle intermetallic phases in the material transition zone, which can occur due to the different material chemical compositions and the related increased risk of fatigue cracking during use.

Advantages of brazing are

- the possibility to join different base materials. As long as the brazing material enables sufficient bonding between each joining partner, multi-material joints can be achieved. This also includes the joining of metal/non-metal parts,
- the low heat input and the related low distortion, which lead to little or no need for correction measures to release stresses or geometry corrections,
- the low oxidation of the joining surfaces due to comparably low temperatures and minimum time to enable oxidation at high temperatures,
- the low line energy, showing a minimum energy input and high process efficiency, and
- the possibly good electrical conductivity of the connection due to metallurgical bonding.

3.7 LASER BEAM WELDING

Laser beam welding is a well-established joining technology, which is largely used in industry due to its fast and precise processing. It includes basically two modes, namely the heat conduction mode and the deep penetration or keyhole welding mode. The laser beam is used as heat source while moving relative to the specimen. In general, the local heat input at high intensity enables a lower energy input compared to conventional welding heat sources and therefore faster processing and decreased thermal impact.

Laser beam welding has many applications including the industrial ones, reaching from thin sheet welding (electronics, micro-technologies, etc.) to tooling or automotive industry, where nowadays more than 60 m laser welding is applied per vehicle, e.g., Volkswagen AG (Wolfsburg, Germany) installed the first hybrid Nd:YAG laser/ MIG welder for aluminum door parts in 2002.

In general, most metals can be laser-welded, while steel, aluminum, copper, and titanium are the most typical materials used. Joining different materials, in particular, dissimilar materials with different properties, can lead to challenges, e.g., regarding the formation of brittle intermetallic phases and increased risk of delamination due to inhomogeneous deformation.

3.7.1 HEAT CONDUCTION WELDING

3.7.1.1 Process Characteristics

Most conventional thermal welding methods aim to heat up the surface of the joining partners to create a melt pool. The incorporated energy is transferred by heat conduction from the surface into the material. The shapes of the melt pools are usually wide at comparably small depths. The ratio of the depth to the width of the melt pools is typically much smaller than one. In conventional welding, the heat-affected zones (HAZs) can become large. The comparably high heat input in the material can induce heat effects around the molten material and affect the microstructure of the heat-treated volume. Most materials experience phase transformations and micro-structural changes at temperatures far below the melting temperature. Since the heat source is placed on the material's surface, volume elements in the melt pool and HAZ experience different thermal cycles. Depending on the locally induced thermal cycle, different phase changes or diffusion processes can be induced. Therefore, besides the weld seam characteristics, the properties of the HAZ must be always considered as part of the welded zone in order to fully describe the joint characteristics.

Such a heat conduction-based welding mode can also be realized by a laser beam as a heat source. The process is called "laser heat conduction welding". Thereby, the laser beam is directed and absorbed on the flat surface of the joining partners' materials. Laser energy absorption takes place on the material's surface partly on solid and/or liquid material depending on the process speed. At low process speed, the laser beam is mostly absorbed by the liquid surface, since the energy input is sufficient to form a large melt pool that reaches large enough dimension to pre-heat the material in front of the laser beam. At very high processing speeds, the laser beam can move faster than the melt pool can form, and the laser beam can be partially absorbed by the solid material. Due to the laser beam illuminating a flat surface, a single laser beam ray is absorbed only one time. In order to avoid back reflections of the laser beam into the optical system, often the laser optic is slightly tilted by a few degrees. The reflected portion of the laser beam from the processing zone is scattered into the surrounding space and is lost for the processing. This means that according to the Fresnel equations, based on the laser wavelength, material parameters, and the surface and temperature conditions, the total energy transfer from the laser beam into the material can be comparably low (e.g., at 1 μm wavelength, for steel ~30%, aluminum ~10% and copper only ~5%). A high reflectivity of the material leads to the necessity that reflective material such as copper requires higher initial laser power (at otherwise constant conditions) in order to generate a melt pool and enable welding. In general, decreasing the laser wavelength increases the absorption (see Section 3.5.2).

Precautions must be taken to avoid unwanted heating or destruction of equipment in the surrounding area of the processing, since the reflected energy can still reach high values. Personal safety equipment is required to avoid injuries, in case there is no protection wall or housing around the processing area.

The absorbed energy on the material's surface is transferred into the material by mainly heat conduction. A quasi-static condition is reached during welding at a constant process speed, which implies that isotherms form at constant distances from the surface heat source. In a cross-sectional view, virtually cutting the material at the position of the heat source, the temperature isotherms define the shape of the melt pool and HAZ. At the melting temperature, the melt pool ends and the HAZ begins.

Typically, a high laser beam intensity is used for achieving the local heating and melting of the joining partners to form a melt pool in both materials. In order to achieve the required high intensities, often the laser beam waist (focal position) is typically positioned on the material's surface or slightly below. The induced melt pool dimensions are therefore relatively small due to the small laser spot sizes. Due to comparably small melt pool dimensions, a large possible gap between the joining partners for making a butt joint must be avoided using suitable clamping and proper edge preparation. In addition, during welding, deformations can occur, which need to be considered when designing the clamping setup to avoid an opening of the gap and welding failures.

The laser beam is used to heat up the material above the melting temperature of at least one joining partner. In heat conduction mode, heating is aimed to induce temperatures below the vaporization temperature of the material. This means that a negligible amount of vaporization losses occur. However, fume and vapor from low-melting alloying elements and remaining dirt or oil on the surfaces can still be created. These emissions require protection of optics using protection windows and cross jets. In certain cases, it may be necessary to install additional gas streams to avoid interactions of laser beam with the produced vapor to avoid absorption losses in the vapor and a resulting reduced energy input into the material.

The heat conduction welding process is typically quite stable. It shows moderate melt convection and can therefore lead to smooth surfaces, low number of melt ejections and achieve high-quality weld seam appearances that fulfill the requirements for customer-visible applications.

Due to the low and local heat input, distortion is very low and little or no post-treatments for geometrical corrections are necessary. The relatively fast processing of 1–200 m/min can reduce the processing costs compared to those of conventional processes, and the short thermal cycles can help to avoid the creation of extensive HAZ and the related unwanted phase creations. In order to avoid effects related to chemical reactions of the molten material with ambient gases (e.g., oxidation), often a shielding gas is applied to cover the processing zone with non-reactive or low-reactive, non-soluble gases, e.g., argon, nitrogen, and helium.

The heat conduction welding can be done with or without applying filler material. It is typically aimed not to use additional welding wire to utilize the fast-processing possibilities and direction-independent remote welding. An additional setup for wire application with a lateral wire feeding system would reduce the processing speed and path freedom. However, for some applications, it may be necessary to add additional material to fill gaps or for altering the chemical composition of the weld seam. Beam shaping guiding the laser beam within the processing optics around a central wire feeding system enables coaxial wire feeding also.

3.7.1.2 Melt Pool Appearance

In general, during heat conduction welding, an energy balance occurs equating the input energy from the laser beam to the losses due to thermal radiation, heat conduction, vaporizing material, temperature increase, and melting (latent heat). Typically, thermal radiation and vaporization effects are small compared to heat conduction effects in heat conduction welding mode.

Assuming the creation of the melt pool by only heat conduction of a surface heat source, a typical model [Röm10] can be used to estimate the depth of the melt pool z

$$z = \frac{2 \cdot A}{4.31 \cdot \rho \cdot c_p \cdot T_m \cdot \sqrt{\lambda_{th}}} \cdot \frac{P_L}{\sqrt{b \cdot v_s}}$$

involving absorption A, density ρ, heat capacity c_p, melting temperature T_m, thermal conductivity λ_{th}, laser power P_L, width of the weld seam b including the HAZ and welding velocity v_s.

General tendencies are:

- A material with high heat conductivity creates a smaller melt pool due to energy losses into the material from the processing zone.
- A higher thermal diffusivity $a = \lambda/c_p \cdot \rho$ leads to a higher necessary energy input to reach the same melt pool dimensions.

3.7.1.3 Melt Pool Movement

In addition to the energy transport by heat conduction, melt convection determines the weld seam geometry in the heat conduction welding mode. Convection transports energy and material, and can therefore alter the processing and geometrical dimensions of the melt pools. Melt pool movement in heat conduction welding is mainly initiated by density differences, the surface-tension-driven Marangoni flow, shielding gas application, and with typically minor impact also by electromagnetic forces, e.g., silicon in aluminum.

Melt flow is induced when density differences occur in the melt pool. Melt pools typically show highest temperatures in the center on its surface with a temperature gradient to the melting temperature at the melt pool border. Therefore, the molten material shows density gradients accordingly. Volumes with smaller density (lower temperatures) experience buoyancy forces and a tendency to upward movement in the melt pool.

In addition, surface tension is temperature-dependent. Therefore, different local temperatures in the melt pool lead to different values of surface tension. When the thermal energy of atoms increases at higher temperatures, in general, surface energy/surface tension is reduced. An area of lower surface tension tries to expand, which induces a melt flow from a high to a low temperature region. This effect is called the Marangoni flow. The effect depends on the properties of a material, in particular the temperature dependence of surface tension, chemical phases, and alloy compositions. The effects of surface tension can be influenced by the surrounding gas impacting the temperature and surface conditions, e.g., by oxidation. Created oxides can slow down the flow and change the local surface tension values. Contaminations (e.g., addition of sulfur to steels) can alter the temperature-dependence of surface tension and even turn around the melt flow under certain conditions (Figure 3.17).

In the moving direction of the laser beam, the melt pool shows similar tendencies that are influenced by the Marangoni flow. Those flow fields can lead to slight bulging effects in the front and rear part of the melt pool (Figure 3.18). Due to the process velocity, the melt pool appears to be elongated toward the rear. This asymmetric appearance can induce complex melt flows and eddies that impact the weld seam geometry and process dynamics.

The effect of temperature gradients on the melt pool can be reduced, e.g., by preheating of the base plate. The intention is to decrease the thermal gradient between

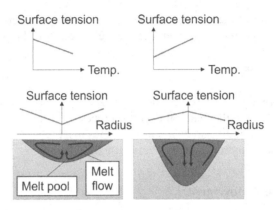

FIGURE 3.17   Surface tension dependence of the Marangoni flow.

the melt pool and the base plate. Thereby, the base plate receives a higher temperature, which is closer to the peak temperature in the melt pool. The smaller thermal
gradients lead to less heat transfer in the melt pool and thereby melt flow is reduced.

Depending on the length of the weld seam, intrinsic pre-heating can occur. The
heat from the processing continuously heats the base plate by the induced energy of
the laser beam (if no active cooling devises are installed). On the one hand, such pre-
heating conditions can stabilize the process. On the other hand, varying pre-heating
conditions can alter the process. This can occur when welding parts and heat accumulation occurs at edges or varying material thicknesses that can conduct the heat in
different ways. This effect can also be the reason for the widening of welding tracks
during the welding process. Countermeasures can be needed, such as a control of
the laser power or active and controlled cooling, in order to avoid process variations.

Typical applications of heat conduction welding can be found in welding of precise features that demand high surface finish, thin-walled components, hermetical
closing of structures, and multi-material joining. The heat conduction melting of
surfaces (without the aim of joining two materials) is also used for re-melting, which
is a form of surface treatment by changing surface properties without additional

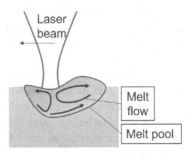

FIGURE 3.18   Sketch of a transversal cross-sectional view of a heat conduction welding
melt pool including melt flows.

material. A side-effect of re-melting can be an increased hardness of the track, while cracks can be avoided by pre-heating (e.g., induction). The method can also be used for repair of cracks, surface smoothening, or increasing mechanical properties.

3.7.2 LASER DEEP PENETRATION WELDING

Modern laser systems enable very high intensity values on the material's surface. Such local high intensities can even induce vaporization. When used for welding, this mode is called "laser deep penetration welding". The total energy input increases and weld seams with depth to width ratios above one can be achieved. Conventional welding methods cannot reach such high local intensities. Electron beams and partly plasma setups can achieve a similar effect of local vaporization and deep impacts inside the material.

3.7.2.1 Deep Penetration Threshold

At increasing laser beam intensity by using higher laser power or smaller laser spot sizes, the energy absorption on the surface initially shows only a slight increase. In the heat conduction mode, absorption mainly increases due to higher absorptivity of the material's surface at higher temperatures, while still only one absorption process happens for each incoming ray (Figure 3.19). However, at the deep penetration intensity threshold I_{th}, a significant increase in absorption can be observed. At this threshold, a vapor channel inside the melt pool is created. The vapor channel occurs when the local intensity is high enough to induce vaporization and create an indentation and then a vapor hole. The vapor channel is also called keyhole, and is typical for the deep penetration process. After its creation, the keyhole is (in optimal case) kept open during the processing and is surrounded by the melt pool, while the laser beam and work piece are moved relative to each other to create the weld seam.

The shape of the keyhole leads to an increased absorptance of laser light. The incoming rays experience multiple reflections and absorption on the keyhole walls

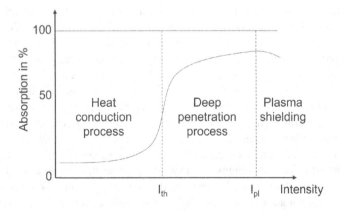

FIGURE 3.19 Absorptance depending on laser intensity including deep penetration I_{th} and plasma threshold I_{pl}.

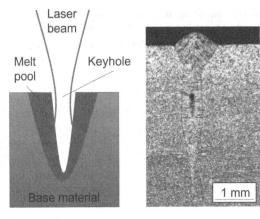

Laser   YLR 1000-SM
Laser power   1 kW
Material   1.4301
Speed   1 m/min
Focal position   -1 mm

FIGURE 3.20 Principle sketch of the laser deep penetration welding (left) and a cross-sectional view of a typical weld seam.

before the rays exit or are completely absorbed [Dau95]. Absorption values of up to 80% can be achieved [Fab05]. This special process feature enables the energy input deep into the material and thereby the deep weld seams.

At even higher intensities, ionized vapor or plasma can form, which can shield the laser beam from transferring energy to the processing zone. Laser energy is thereby absorbed to ionize vapor or gas, which can decrease the total absorption into the work piece.

In the deep penetration mode, laser energy is transported and absorbed into deeper regions of the material. Therefore, typical deep penetration weld seams show a comparably large, melted volume and the typical deeper than wide cross-sectional shapes. Due to significantly higher energy input, the weld seam cross sections are typically also wider than those in conduction mode. Figure 3.20 shows the schematic sketch of a laser deep penetration process and a typical weld seam cross section, with a pore in the center of the weld seam.

The intensity threshold to reach deep penetration welding modes mainly depends on the properties of a material and wavelength of laser. When using CO_2-lasers at 10.6 μm wavelength, the threshold is in the range of $1–2 \times 10^6$ W/cm². At wavelengths around 1 μm, the threshold can already appear in the range of 10^4 W/cm². On the material's side, a high heat conductivity of the material leads, in general, to less available energy for the process due to energy losses into the bulk material, which leads to a higher required threshold to reach deep penetration. Aluminum or copper, with their comparably high heat conductivity, show this effect. Also, the absolute value of the boiling temperature and the difference between melting and boiling temperatures as well as latent heats of fusion and vaporization can influence the total absorption and thereby the intensity threshold. In general, a higher absorption of the material at the used wavelength increases the energy input and therefore can lower the intensity threshold. For welding, the relative movement of the specimen and the laser beam affects the energy input in addition. A low line energy (low laser power or fast welding speed) leads to a decreased energy input and even at high beam intensities, the deep penetration threshold might not be reached.

The most significant thermal properties of materials impacting the total absorption are:

- Absorption coefficient,
- Thermal conductivity (W/(m·K)),
- Heat of fusion (J/g),
- Specific heat capacity (J/(kg·K)),
- Heat of vaporization (J/g), and
- Heat of phase transformation (J/g).

3.7.2.2 Parameter Impact

The laser welding process can be modified by several process parameters to adjust the energy input and design the weld seam shapes. Figure 3.21 shows typical tendencies when welding at different processing parameters with increasing molten volume at lower processing speeds, higher laser power, and a focal position slightly below the material's surface [Web10]. A higher reflection (lower absorption) is expected at higher processing speeds due to an increased illumination of flat, cold material.

In practice, it is advised not to weld in the transition zone around the deep penetration threshold to avoid instabilities due to varying welding modes. Already slight absorption variations can lead to the shift from heat conduction to deep penetration mode within one weld seam.

Although the word "deep" is in the mode description, the main characteristic for distinguishing the welding modes is the reaching of vaporization. Therefore, the deep penetration mode can be achieved in foils of ~20 μm thickness as well as in sheets of >30 mm thickness. Micro-welding applications can result in small weld seam depths but can still be produced in the keyhole mode.

Since energy input is increased in the deep penetration mode due to multiple reflections and absorption, the process is more energy-efficient compared to heat conduction processes, where most of the laser beam energy is reflected. In addition, the process enables welding of materials that are considered difficult to be welded and material combinations of multi- or even dissimilar materials.

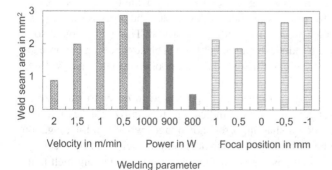

FIGURE 3.21 Impact of processing parameters on weld seam areas in cross sections at varying process parameters from the reference (ref) parameters.

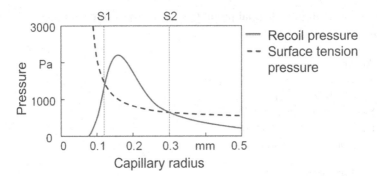

FIGURE 3.22   Pressure balance inside the keyhole [Vol16a].

3.7.2.3   Keyhole Energy and Pressure Balances

Inside a stable keyhole, the energy input from the laser beam by multiple reflections and the related multiple absorption inside the vapor channel is balanced by energy losses. Energy losses mainly contain heat conduction into the bulk material and emission of vapor (e.g., [Vol16a]). Typically, melt pool convection and thermal radiation play a minor role in deep penetration welding.

The pressure balance consists of the two main components of the laser-induced recoil pressure due to vaporization counteracting the surface tension pressure of the surrounding melt that tries to close the keyhole. The permanent illumination of the keyhole walls induces vaporization and the increased pressure inside the keyhole that keeps it open. The created vapor recoil pressure and the increased pressure push the keyhole walls open acting against the surface tension pressure (Figure 3.22).

Two mathematically possible stable conditions can be found (S1 and S2, Figure 3.22), while only S2 is physically stable, which is the local keyhole radius appearing when no other disturbances appear. It was shown, however, that keyholes are intrinsically dynamic changing depths and widths at high frequencies, even in cw laser mode, e.g., in high-speed X-ray measurements (e.g., [Hei13, Kat02]).

3.7.2.4   Beam Shaping Opportunities

Although the deep penetration welding mode offers the advantage of creating deep and narrow weld seams, the process can become unstable. The process stability cannot be guaranteed yet for all processing conditions [Bro08]. The interaction of solid, liquid, and vapor phases can induce dynamic behavior. The keyhole can collapse and entrap surrounding gas, which can form pores, melt pool ejections (e.g., [Vol18a]), or cracks.

Therefore, beam shaping approaches are largely used in order to stabilize the process. Multi-beam arrangements showed indications to stabilize the keyhole (e.g., [Vol19a]). Also, dynamic beam shaping offers advantages when welding, e.g., aluminum (e.g., [Pri20]). The stabilization of keyholes can be achieved by creating a wider keyhole opening. This helps the vapor to escape without ejecting melt material. Ring beams can achieve the energy input to create a wider keyhole opening (e.g. [Wan22]).

A typical challenge for laser welding processes is the appearance of a gap between the joining partners. Due to comparably small dimensions of the laser beam, a gap can lead to a partial or even complete laser beam transmission through the gap without energy transfer to the material. In order to still use the advantage of a local high-intensity energy input and avoid thermal effects like extensive distortion, beam shaping can be used to guide the energy to the locations where needed. For enabling gap bridging, the energy input is required to be transferred to the joining partners. Therefore, the distribution of laser beam energy to two or even more single spots can be advantageous [Vol23b]. The initiated melt pool flows can even support the closing of the gap with the available melt. In addition, rapid beam movements induced by deformable mirrors can support the bridging of the gap [Gug23].

3.7.2.5   Energy Absorption in the Keyhole

The keyhole geometry is a result of the energy and impulse balance. The photonic energy from the laser beam is absorbed multiple times due to multiple reflections on the keyhole walls, which denotes the energy input. The main energy losses from the keyhole walls are the heat conduction into the melt pool and then into surrounding bulk material and the convection in the melt pool with a typical minor loss due to thermal radiation. In many cases, mainly the front of the keyhole wall is directly illuminated by the laser beam, which can result in downward movement of melt on the front wall by the induced recoil pressure. The rear wall is pushed backward by the extensive vapor produced on the keyhole walls and the related increased gas pressure inside the keyhole. In addition, recoil pressure is induced due to multiple reflections of laser light to the rear, side, and front keyhole walls. However, the keyhole can be unstable and can even collapse during continuous illumination. There are several possible reasons for these instabilities:

1. The keyhole front is inclined due to the energy distribution in depth. Therefore, the energy input into the lower keyhole sections is low. In addition, the process velocity defines the keyhole inclination, while a higher process velocity leads in general to higher inclination angles. The angled keyhole front and the induced wavy keyhole front surface can lead to varying absorption and in particular reflection angles. Reflected beams can strike the rear wall at different positions over time, which can induce keyhole rear wall fluctuations. A second effect is the angle-dependence of the Fresnel absorption, which denotes that at different keyhole wall angles, absorption can vary and thereby the local ablation pressure.

2. Vapor and plasma absorption effects can occur that can temporarily absorb parts of the laser light before being used for absorption in the keyhole and thereby induce depth and radial keyhole fluctuations.

3. Even at continuous energy input, the keyhole shows intrinsic fluctuations. Natural spiking has been observed [Bol13], denoting high-frequency depth variations, which, however, not necessarily result in bad weld quality.

4. Melt pool induced melt pool waves can transport melt toward the keyhole, which can induce and support temporary closing of the keyhole. This phenomenon was mainly observed to close the keyhole's top section partly or

completely and in the middle sections of the keyhole [Fab10], disconnecting the lower keyhole part from laser irradiation. Those collapses are often harder to identify with visualization methods and can entrap gas, which can form pores.

5. In an industrial environment, impacts of vaporizing dirt or inclusions can also lead to process discontinuities and quality reduction.
6. Material-specific characteristics can lead to energy absorption variations such as created oxide layers that show different absorptivity than the base material. This effect is pronounced when processing, e.g., aluminum alloys, where the high-melting oxide layer can change the surface of the melt pool geometry and melt pool flows as well as the related melt pool waves, laser beam energy absorption, and even hinder gas bubble escape from the melt pool.

3.7.2.6 Plasma Absorption

Besides the absorption of laser energy by the materials of the joining partners, plasma can be formed during laser welding at high intensities (cw at ~10^9 W/cm^2) in free space, which denotes the upper limit of manufacturing with lasers. In dirty environments, the intensity threshold can be even lower.

On the one hand, the absorbed energy in the plasma is not transferred to the process zone directly, which hinders the establishment of the keyhole and the desired deep penetration process. On the other hand, the energy in the plasma can be further radially transported to the keyhole walls when the plasma burns inside the keyhole and increases the energy input that way to widen the keyhole.

Plasma develops due to the "inverse Bremsstrahlung", which occurs when a photon is absorbed by an electron during collision with a nucleus and thereby directly accelerates ions. This kind of absorption can be very efficient. Hot gas (surrounding gas or produced vapor during the process) is ionized by the photonic energy and forms plasma when the material-dependent threshold ionization energy is overcome. In general, plasma formation is more likely at high ambient pressure, higher laser wavelength, and lower ionization potential of the gas/vapor. The presence of free electrons makes the plasma conductive, but neutral to the outside. Plasmas show very high temperatures and can change the refractive index, which can result in different path lengths of the laser beam rays in plasma. As a result, increased beam spot sizes, a related lower intensity, and defocusing occur. This can impact the laser processes. Often the transferred laser energy is reduced or fluctuates.

Plasma absorption is (in a simplified estimation) proportional to the (laser wavelength)^2, which means that the ~10 times higher wavelength of a CO_2 laser (10.6 μm) compared to a typical solid-state laser (~1 μm), plasma absorption is ~100 times more efficient [Zel66]. In turn, this also means that plasma formation is less likely when using 1 μm lasers, reducing the plasma shielding effects, and shielding gas selection becomes less significant in this respect.

3.7.2.7 Vapor Outflow

While initiating vaporization of a surface by a laser beam or also during pulsed processes, a shock wave is created, which initiates a rapid build-up of a high vapor pressure on the surface. When the vapor expands from this zone, the surrounding gas is pushed away, and a recoil pressure is created that indents the melt pool. This leads to a geometrical deformation and the creation of the keyhole when the intensity of the laser beam is high enough. A special non-equilibrium state on the surface is created, since more particles leave the surface than condensate. This thermo-dynamical imbalance in the near-surface region is called the Knudsen layer, where both pressure and temperature jumps occur (e.g., [Fin90]).

The vapor of high pressure created inside the keyhole by laser ablation needs to escape the keyhole. If the keyhole is only open toward the laser impact direction, the process is called partial penetration and the exiting vapor fully expands into the upward direction toward the laser beam. At full penetration, the keyhole has an additional opening on the lower part of the sheet, which enables the vapor to escape in two directions. Typically, full penetration processes are intrinsically more stable.

When the vapor is created on the keyhole walls, it exerts the recoil pressure on the wall and exits perpendicularly from the local wall surface. Depending on the inclination of the keyhole wall, the vapor can be pushed out of a keyhole opening or impact other parts of the keyhole and deform those elements. However, when the vapor tries to exit the keyhole, there are shear forces created on the keyhole walls when the vapor passes. These can counteract the typical downward movement of the melt pool waves on the keyhole front and lead to complex surface forces on the vaporizing melt.

The exiting vapor can reach speeds of 100–200 m/s, which is why in practice, a cross jet (gas jet horizontally applied) is used below the optics to protect them from contamination. The mass loss has been shown to be minor compared to the typical melt pool volumes ([Bec96, Dul99]). At smaller keyholes, the probability that the vapor produced at the keyhole front condensates on the rear keyhole wall is higher and reduces the mass losses in general. The ejected particles have been shown to be in nanometer dimensions (aluminum: 200 nm; steel: 70 nm, [Vol18a]), which condensate and slow down after exiting the keyhole opening(s). The small particles can lead to laser beam scattering and absorption effects [Hul57]. If the particle size is much smaller than the wavelength, Rayleigh scattering occurs, which means that small particles are forced to oscillate with the frequency of the incident light and emit radiation, which is visible as scattering. If the size of particles is in the range of the wavelength or larger, Mie theory can be applied to describe the scattering. In contrast to intuition, larger particles allow, in general, more energy propagation in laser axis direction. For particles much larger than the wavelength, geometrical optics calculations can be applied. Additional effects can occur when dirt or water vaporize that are not fully removed before processing. Similar effects are seen when coatings or claddings are present on materials.

3.7.2.8 Melt Flow

In the melt pool, similar to the energy balance, the mass balance must be fulfilled, which is relevant when describing the melt pool behavior during laser deep penetration process. In addition to the temperature-dependent, surface-tension-driven Marangoni flow, in the deep penetration welding mode, additional melt flows are induced due to the formation of the keyhole. The keyhole is a vapor channel and is

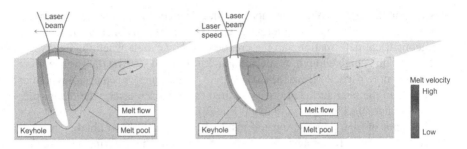

FIGURE 3.23 Qualitative sketches of melt pool dimensions and melt flow directions, and velocities at low (left) and high (right) processing speed.

therefore an obstacle for the melt flow (Figure 3.21). On the keyhole front side, the melt pool is typically comparably thin due to the high processing speed and is pushed downward by the recoil pressure. The keyhole front has been seen to show a wavy structure [Kap13]. It has been shown that the keyhole front waves move downward by the increased local absorption on the waves, which induces a locally high recoil pressure. This additional effect increases the velocity of the melt and can induce additional flow patterns. This effect is complex to mathematically describe and is counteracting the impact of shear forces due to the exiting vapor, which mainly pushes the melt upward.

In addition, the melt created at the front of the keyhole must move around the keyhole when the laser beam progresses. Since the keyhole contains no melt, it is an obstacle for the melt movement. Due to the long travel distance around the keyhole instead of taking a straight path, the melt is accelerated (Figure 3.23). The reached velocities can be much higher than melt pool velocities observed in heat conduction processes. The melt moves in the typically thin melt film around the keyhole. In the larger melt pool behind the keyhole, the melt can decelerate and induce complex melt pool movements as well. Due to the large differences between the melt pool velocities from the keyhole region and the region behind the keyhole, complex eddies are typically formed. Local high melt pool velocities or accelerations can lead to melt pool instabilities. These can be caused by surface waves that combine to larger waves or impulses inside the melt pool pushing melt material in different directions. These instabilities can induce melt movement on the surface that can overcome the surface tension and create melt pool ejections.

In order to influence the melt pool appearance, several strategies can be applied. In general, more space for the melt pool to move can stabilize the process. A higher material heat conductivity or a slower processing speed can increase its sizes and stabilize the processes. Also beam shaping approaches can support the widening of the melt pool in two possible ways. On the one hand, the melt pool can be directly illuminated by additional heat sources to increase its width, e.g., using a second defocused laser beam. On the other hand, the energy input into the keyhole can be varied, e.g., by increasing the local intensity on the keyhole side walls to widen the energy input locations. It is also possible to actively influence the melt pool movement, e.g., by applying volume or surface forces by using gravity or electro-magnetic stabilization [Lin01]. Lorentz forces can also be active, e.g., during arc or hybrid processes. Gravity forces can influence the melt dimensions and flows, in particular, at larger melt pools.

The melt pool must, however, be seen as a system. When altering the melt pool dimensions, the mass and energy transport are changed as well [Miy03]. Depending on the created melt pool flows, the melt pool mixing is changed. This can possibly lead to inhomogeneous element distribution in the weld seam, when mixing is not sufficiently induced. In addition, the melt pool movement and related conductive energy transport define the local cooling rates and thermal gradients. They, in turn, define the formation of cracks and also the development of microstructures. In addition, the induced temperature distributions in and on the melt pool will induce different melt flows and also different shear forces from vapor and gas flows. Material properties such as surface tension counteracting the spatter ejection formation, are defined by the melt pool temperature and therefore can impact the melt pool dynamics as well.

Melt flows on the melt pool surface can be observed by visual monitoring (e.g., high-speed-cameras). Many flow features can be observed by visual monitoring such as the Marangoni flow patterns, oxide movements, or spatter ejections. This information can often help to validate numerical models already.

However, the melt flows inside the melt pool cannot be directly observed by conventional visual methods. There is a possibility of using X-ray observation techniques to learn about the melt movement inside a melt pool. By placing high-melting particles (e.g., tungsten) as tracer particles in material, their travel paths and speeds can be observed. While 2D observations give indications of the melt pool movement [Rei21], 3D systems are used to get a full picture of the geometrical movement of the melt pool flows. Another possible observation technique is the welding behind high melting glass. Although the properties of glass will not fully represent regular welding process conditions regarding heat conductivity or absorption, melt flow patterns can be observed (e.g., [Hua18]).

Element losses due to vaporization can be an additional factor to locally change material properties (e.g., surface tension or viscosity), which can alter the melt flow. In addition to the recoil pressure induced by the laser beam, the exiting vapor can induce friction forces on the keyhole walls, which can induce melt pool momentum.

3.7.2.9   Gas Flows

Most laser processes are conducted with a surrounding gas atmosphere at ambient pressure and temperature. Vacuum is seldom used due to the additional effort to clamp parts in the pressure chamber and evacuate the chamber before processing. However, it has been shown that some advantages can be observed when welding with reduced ambient pressure. Since a reduced amount of gas is available, porosity from incorporated gas is reduced (e.g., [Jia19]). In addition, the vapor plume has been seen to exit more straight and less dynamic from the keyhole opening, leading to a calmer process (e.g., [Ols16]).

However, one advantage of laser beam processing compared to, e.g., electron beam processing is that vacuum is not necessary to achieve processing, which makes part handling and processing, in general, much easier. However, it means that the gas around the processing zone is an influencing factor on the process results. Both the kind of gas and the gas flows must be considered an interacting medium during welding. Due to comparably low energy transfer between the melt pool and the surrounding gas, typically the energy loss by gas convection can be neglected. However, a directed gas flow can decrease the surface temperature of the melt pool. This effect can be desired to induce, e.g., rapid cooling or an additional heat transport to enable higher energy inputs. In addition, the local variation in surface temperatures can influence the surface tension and lead to additional Marangoni flow effects. The melt flow can in turn affect the melt pool geometry development.

Often, shielding gases are applied through gas nozzles to cover the process area. The nozzle outlets are either round- or oval-shaped, with a variety of variations to optimize the gas flow tailored to the process zone size and the purpose of the gas application. Typically, the gas flow rates operate at moderate flow rates. Too low flow rates can lead to insufficient coverage of the process zone, while too high flow rates can induce turbulent flows and eddies that can lead to a gas mixing with the ambient atmosphere and therefore to unwanted gases in the processing zone. In addition, high gas flow rates can impact the shape of the molten surfaces. The inertia can indent the melt pool and change its shape and flow fields, which is typically not desired.

Typical welding gases used are argon, nitrogen, or helium. The purpose of gas application is mainly the separation of oxygen in the ambient air from the melt pool to avoid surface oxidation. Therefore, low reactive and inert gases are often used. However, some gases alter the chemical composition in some materials, e.g., steels form nitrides when using nitrogen as shielding gas. Such in-situ alloying effects are typically not desired and other alternative gases are used. However, it is possible to tailor the chemical composition that way and alter the material during processing. Then, the gas even defines the metallurgy and, in the end, the mechanical properties of the welded zone. This approach is not widely used due to the complex and partly unpredictable amount and location of the reactions. In addition, the mixing of the new phases in the melt pool is difficult to control and can form undesired microstructural features or brittle phases.

The kind of gas used is of particular importance when ionization and plasma formation are possible. The ionization possibility of the gas between the laser optic and the processing zone defines (1) if ionization or plasma formation occurs and (2) at which laser energy levels the laser energy absorption is high enough to initiate the ionization of the gas.

In addition, there is a metal vapor outflow from the keyhole that mixes with the shielding and surrounding gas. On the one hand, the additional vapor flow can induce complex gas flows when interacting with the shielding gas flow and lead to turbulent flows. On the other hand, metal vapor can also form plasma. This can either even occur inside the keyhole, where the vapor pressure is high and many metal particles are present or above the processing zone. Plasma formation is most relevant to consider when using large wavelengths, e.g., 10.6 μm produced by CO_2 lasers. Plasma can lead to scattering of the laser beam, which can decrease the transferred laser beam energy to the processing zone and increase the beam diameter on the material, denoting a reduction in the local intensity. Plasma energy absorption can lead to partial or even full transfer of the laser energy by plasma, reducing the energy coupling into the material. The decreased power and intensity can interrupt the process or change welding modes from keyhole to heat conduction welding.

An additional gas flow that is typically applied is the cross jet. The cross jet is installed below the laser optics to protect them from vapor and spatters moving toward the optics. The cross jet is a high-pressure gas flow perpendicular to the laser beam axis. This high-velocity gas flow can also impact the gas flow field around the processing zone and might interrupt the low-pressure shielding gas flow. This effect is more pronounced when the focal distance of the focusing lens is short, and the laser optics must be positioned close to the processing zone. When testing new processes or running highly dynamic processes, it is often necessary to install additional cross jets.

3.7.2.10 Solidification

The solidification process in welding is, in general, similar to the ones during casting processes, but some additional effects can occur. In welding, the cooling of the process zone happens due to the self-quenching effect since the heat source moves further from the previous heat input position and heat is further conducted into the base material. In many thermal welding processes, the cooling rates can be moderate leading to moderate thermal gradients and solidification comparable to casting effects. However, in laser welding, high thermal gradients from either the overheated melt pool or the keyhole process having temperatures at or even slightly above the metal boiling temperature to the melting line are present.

Due to the relative movement of the laser beam and the work piece, the melt pool is typically elongated toward its rear part. This also means that the solidification front on the rear and side sections of the melt pool can be different. The typical fast cooling during laser treatments leads to, in general, a comparably increased grain growth speed since there are many crystallizing centers on the melt line. Grain growth is often seen to be dendritic but cellular growth can also be possible. The direction of crystal growth is usually parallel to the heat flow (Figure 3.24). The diffusion speed and the nucleating agents are therefore inhomogeneous, and

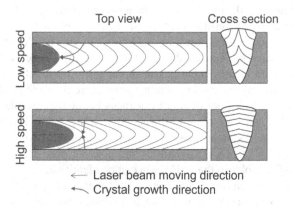

FIGURE 3.24 General trends of grain growth directions during laser welding at different processing speeds.

crystallization starts at certain grains or contaminations, which increase the risk of cracking. In order to achieve fine grains, grain refining elements can be added to the alloy.

3.7.2.11   Weld Monitoring

Monitoring of measuring equipment and the laser process during welding is an important part of the processing to guarantee continuous reliable process conditions and weld quality. Laser power is typically observed by the laser machine itself and time-resolved protocols can be extracted for documentation or the search for origins of failures. External power measuring systems can be used to identify how much power is actually transferred to the processing zone. This is, in particular, important to identify potential power losses in the optical system from the laser machine to the processing zone. Such power measurements can be done during production cycles to check for power variations. This is recommended to perform on a regular basis to ensure that, e.g., contaminations like spatters on the protection window are identified and do not lead to equipment damage or part failures.

Apart from power measurements, it is recommended to perform caustic measurements to ensure that the laser beam shows the expected profile. In particular, the focus position must be correctly positioned for welding processes. Caustic characteristics are possible to change when the optical system is contaminated, or optics are misaligned. In addition, beam measurements can be used to verify the correct spatial intensity distribution in the working plane.

For detecting the gap between the joining partners, it is possible to mount laser triangulation sensors to the laser optics that identify the gap position and give feedback to the positioning machine/robot to adjust the laser movement to ensure the correct laser position for welding. Laser triangulation sensors measure the surface topology along a line, which offers the possibility to use it for even post-measurements. For that, the triangulation sensor is mounted to measure the geometric shape of the produced track. This information can give indicators about the success of the welding process. Even open pores, spatters, or misalignment can be identified.

Optical coherence tomography (OCT) is increasingly used for weld process monitoring. When scanning the measuring beam across a surface, gaps between the joining partners can also be identified and the laser beam positioning is guaranteed. Like triangulation sensors, cracks, pores, or undercuts can also be identified when used as post-process sensor. OCT offers the additional possibility to be used as in-process sensor. The distance measurement has been shown to be feasible inside the keyhole [Web14]. That way, a direct possibility to measure the welding depth is given. Although there is a melt pool of small thickness below the keyhole, the measurements of keyhole depths are close to the actual welding depth. The high recording frequency of the OCT systems enables the detection of the keyhole depth fluctuations as an indicator of process dynamics and possible pore creation. In addition, drop-outs as well as holes in the weld seam can be identified.

Visual cameras are an important, comparably cheap, and robust way of the laser welding process monitoring. Cameras can be mounted externally or even coaxially on the laser optics. Coaxial camera mounts are often provided by laser optic producers. These cameras use the optical path of the laser beam to record the processing zone. Visual cameras are typically used to observe melt pool dimensions and dynamics. Changes of melt pool width or length indicate process inaccuracies or defect creations. Automated analysis tools can help to identify the melt pool shape in the recorded frames and inform the operator at significant deviations. Depending on the amount of data and data transfer speeds, the frame analysis can either be done after or even during the process. The evaluation during the process gives the opportunity to install closed-loop control systems that automatically change process parameters to counteract a recorded tendency. More expensive recording equipment are high-speed cameras. These offer the possibility to identify many processing details and are therefore mostly used in science and research.

Photodiodes are used when the brightness in a certain spectral range indicates process conditions, e.g., photodiodes recording in the IR range can indicate changes in the surface temperature. Often diodes are used to identify the exact laser on/off times or to trigger other equipment accordingly.

Thermal monitoring is a widely used technique for identifying process variations and even defects. Although it seems intuitive that thermal processes are monitored with thermal measurement systems, the derivation of reliable quantitative values is difficult. The main challenge is the variation in emissivity of the material both with temperature and at phase changes. This would require the adaption of emissivity in temperature recordings to the correct state and temperature of the material at every point. However, one-color thermal imaging is used, e.g., to measure solidification characteristics such as the so-called $t_{8/5}$ time. In addition, thermal images can give indications about the weld seam dimensions as well. Two-color thermographic imaging gives the possibility to get emissivity-compensated temperature data as well but requires additional calculations. Thermal imaging is feasible to use to identify process changes. Even when the exact temperatures are not correctly given, temperature changes can be often visualized well. At local overheating, connection problems can be the reason, e.g., false friends or crack occurrence. In both cases, heat conduction into the material is interrupted, leading to a visible increase in surface temperature; even larger pores can be found. Also, fluctuations of the laser power can be potentially seen and identified before process interruptions occur.

X-ray monitoring techniques offer a wide range of observation possibilities during laser processes. During welding, X-ray radiographic monitoring is used to identify keyhole dimensions and melt pool movements. Synchrotron facilities offer the possibility to install scintillators that translate X-ray images into visible images (e.g., [Kau23]). These can be recorded with high-speed cameras, which give new insights into the process behavior. Tracer particles of high-melting materials can be used to track the melt pool flows. In addition, diffractometry measurements are used to identify phase change kinetics in the material during welding (e.g., [Oli16]).

Recent monitoring of the keyhole process using X-ray imaging of laser polishing has revealed some new features of the keyhole behavior [Fau23]. An experiment was designed in which the laser beam polishes a metal surface. However, the surface showed an applied surface structure that the laser beam had to pass. A relatively stable keyhole was created, and the laser beam approached the surface feature, denoting an obstacle. After the surface feature, the keyhole changed shape. It became deeper and thinner without changing the process or material parameters. It is typically assumed that one set of process parameters leads to one particular shape of keyhole. However, the history of the keyhole seems to play an important role as well. It was hypothesized that while the laser beam was illuminating the surface feature, the keyhole dimensions changed and thereby the internal multiple reflections altered as well. Based on this initial shape, which differs from the regular keyhole creation process from a flat surface, the keyhole shape develops in a different way leading to a different keyhole shape. Such knowledge can be used to tailor the process to achieve the desired polishing or welding outcomes.

3.7.2.12 Comparison with Other Welding Processes

Laser beam welding can be used as a complementary welding method to other welding processes. In general, autogenous laser beam welding is used for automated processes, where fast processing, high precision, and little heat input are advantageous. Metal inert gas (MIG) welding, metal active gas (MAG) welding, and tungsten inert gas (TIG) welding are beneficial due to the requirement of little and cheap equipment and reduced effort for protection equipment and personal safety measures. In addition, comparably high heat input into the processing zone easily bridges the gap due to, in general, wider melt pools (Figure 3.25).

Electron beam welding can reach similar dimensions of the keyhole and melt pool sizes compared to the laser beam welding processes. Weld seams produced by electron beam welding show, in general, less porosity and can be more dynamically stable compared to laser weld seams. However, the process requires vacuum, which needs additional effort in part handling due to increased lead times during creation of vacuum in the mandatory chamber. In addition, the dimensions of vacuum chambers can restrict the sizes of parts and components possible to join. Attempts are ongoing to generate a local vacuum to enable the processing of large structures as well.

General advantages and disadvantages of laser welding are summarized in Figure 3.26. The investment in a laser machine can be higher compared to that in equipment for other welding methods. However, in the long run, the investment in lasers can show benefits due to their fast processing possibilities, which reduces

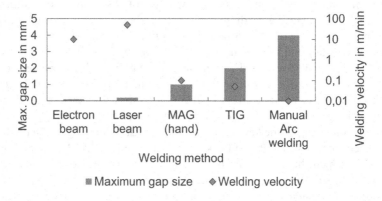

FIGURE 3.25 Comparison of bridgeable gap sizes and typical welding velocities between several welding methods.

Advantages	Disadvantages
• High welding speed • Big depth to width ratio • Low heat input -> small HAZ and less risk of deformations • No mechanical contact to workpiece • No need for filler metal • Easy to automate	• High investment cost • High demand on fixtures and joint preparation (butt joints) • Low heat input can result in cracking • Can not weld high reflective materials • Low electric efficiency

FIGURE 3.26   Laser welding evaluation.

lead times and costs for processing. In addition, the costs for maintenance are comparably low since the laser machines usually have long lifetime. Since the laser beam is a remote tool, there are no direct tool wear issues. However, gas consumption and the regular need to change protection glasses add some maintenance costs and work.

For laser beam welding, joint preparation is often necessary. Due to the small dimensions of laser beam, the gap must be typically small and constant during welding to avoid interruptions or failures. Therefore, an effort into a proper clamping system is needed to keep the material in place. However, the effort can be justified when welding similar parts and components multiple times.

The low heat input is advantageous for avoiding extensive distortion. Laser machine producers continuously improve the energy efficiency of their laser machines to enable a need of minimum energy for processing. In principle, all materials that absorb the laser light of the used laser wavelength can be processed. High-reflective materials absorb less energy of the laser beam and show challenges for processing and low energy efficiency.

The laser shows advantages in the achievable welding speed and possibility to form a keyhole leading to the deep weld seams. The low heat input minimizes the creation of the HAZ and the risk of deformations.

3.7.3   WELDABILITY

Weldability defines how easy or feasible the welding process is in different circumstances and conditions. The main three aspects that influence the weldability are the material properties, the construction/design, and the welding method (Figure 3.27).

Welding method	Material	Construction
- Energy source and input - Shielding gases and feedstock rate - Clamping and process speed	- Absorption/ Reflectivity - Heat conductivity - Chemical composition	- Surface conditions - Joint arrangement - Part geometry

FIGURE 3.27   Impacts on weldability.

The welding method defines weldability due to energy source, process speed, and the related energy input properties. The method also includes the necessity that the welded material is conductive for some welding methods such as in MIG welding. The welding setup as part of the welding method such as shielding gas application, the kind of gas, and the possibly applied additional feedstock material (typically wire) influences the weldability as well. Welding of a certain material might be possible when using one type of gas but not when using another one.

Welding with wire leads to different laser energy input mechanisms. Due to added material, the surfaces appear in a different shape, typically higher melt pools at more spherical shape are found. On the one hand, this will change the energy input due to the changed incident laser ray angles on the surface. On the other hand, melt pool dimensions will change or higher laser energy input is needed to achieve the same weld seam dimensions.

In addition, the clamping has a significant impact on the welding results. Clamping defines the positioning accuracy and gap sizes. In addition, the small, but existing deformation, needs to be compensated by the clamping setup. Clamping induces stresses before welding to maintain the joining partners in place. During welding, stresses can increase due to thermal impact. In the worst case, after removing the component from the clamping device, the residual stresses can lead to deformations of the component or even to cracking. The clamping procedure can therefore exclude welding methods usable for a certain task.

Material properties define the weldability. The material defines if laser energy can be transferred into it. The laser energy input must be possible when, e.g., the absorption of the material must be high enough to initiate laser melting. In addition, heat conductivity, defining the heat transported away from the processing zone, defines the creation of the melt pool dimensions and the HAZ. If too little energy is absorbed or the heat is removed too quickly, welding might not be possible. The chemical composition of the welding material can influence the weldability in many ways. The welding material can be the base material (if the same material is used for both joining partners), the mixing of the base materials (if multi-material joints are made), or the mixing of the base materials with the filler material (e.g., when welding with wire addition). It is possible that elements lead to chemical reactions within the melt pool and create new phases and/or are an additional heat source (at exothermic reaction) or heat sink (at endothermic reactions). In addition, the elements on the liquid surface can react with the ambient gas or integrate as alloying elements or reaction products into the melt pool. This effect can be desired to create certain carbides or nitrides but is usually not wanted. Element vaporization can occur, while elements with low boiling temperatures evaporate first. Those element losses lead to changes in the chemical composition of the melt pool. In addition, local element variations can occur that can lead to diffusion processes and additional melt flows. This can lead to the related induced melt pool dynamics or even induce spatter ejections. The additionally created recoil pressure of the evaporating elements can also induce indentations on the melt pool, with the consequences of a deformed surface with respect to absorption and restrictions or initiation of melt pool flows.

The construction and design of components or parts define the accessibility of tools used for different welding methods. When using methods that require the access of welding zone with a welding torch (e.g., MIG, MAG, TIG) and a wire, it must be possible for the torch to be in a position close to the processing zone including the enabling of the mandatory movement during welding. A laser beam with a good beam quality used at a long focal distance can show advantages related to accessibility. Together with the possibility to automize the process, the accessibility can define how well the construction is suitable for the welding method and how efficiently the path planning can be done.

Additionally, errors can reduce weldability. Three error types can be distinguished. Constant errors do not depend on the number of manufactured pieces but are caused by, e.g., wrong use of tools or clamping. In contrast, systematic errors are known errors, e.g., due to wear of tools and are predictable since they can be calculated from one piece to the other. Random errors (e.g., between different batches, temperature variations, impacts of dirt) are difficult to avoid or predict.

Furthermore, following expressions are defined:

- *Welding suitability*: "Is the material suitable to be used in combination with the intended welding method for this particular part design?" On the material's side, the general tendencies are that more alloying elements lead to worse welding results. Typically, intermetallic phases that can occur lead to brittle phases in the weld seam when using, e.g., steel. Many alloying elements also lead to an increased melt interval, e.g., in aluminum alloys and increase thereby the risk of hot cracking. Alloying elements can form inclusions and lead to more available gases and particles in the vapor, which can lead to more spattering and increased pore formation. In addition, hardening effects are typically more pronounced at higher carbon content in steels, which can reduce the strength of the welded zone, when carbides are formed. As a rough rule of thumb, welding steel with carbon content below 0.22% is suitable (see also Schaeffler Diagram for the impact of alloying elements on weldability).
- *Welding possibility*: "Is it possible to weld the part with the chosen method?" In case the process works for the chosen material or material combination, welding possibility must be given. The accessibility of the joining region must be guaranteed by joint design and the correct clamping or pre-joining. The joining zone should be cleaned from, e.g., oil, water, and oxides and if needed pre-heated in an oven or with the help of another heat source such as a burner, a plasma source, or another laser beam. Further equipment might be needed to track and control the weld position to enable welding. It must also be checked if the necessary post-processing is possible, e.g., if the parts are suitable for being heat treated in a furnace with respect to the required temperatures and geometrical fit of the component in the furnace. In addition, spatter removal must be considered in the post-processing and accessibility to the relevant surfaces with the removal tools need to be guaranteed. Similarly, distortion correction must be possible as well as the release of residual stresses.
- *Weld safety*: "Will the weld quality be achievable with the chosen setup?" The requirements of the welded zones and the HAZ must be fulfilled in

relation to the maximum allowed imperfections and material properties such as strength, hardness, or fatigue life properties. This can require process control, destructive or non-destructive testing, fatigue testing, and material simulations.

3.7.3.1   Weld Seam Quality

The quality of a weld seam is typically evaluated after the welding process. There are a variety of methods for evaluating several aspects of weld quality, while only a few most common methods are mentioned here. Depending on the application, a certain number of parts are taken from one batch to perform quality checks. Parts from the same batch are assumed to show comparable quality features. In critical applications, it is possible that all parts need to be tested.

In general, there are destructive and non-destructive evaluation methods. Destructive testing is necessary to derive certain parameters, whereas non-destructive testing is of course preferred in order not to waste the tested parts. Parts tested with destructive testing methods will not be usable even if they pass the test. Destructive testing is often costly and time-consuming. However, it can be useful for design purposes.

Destructive testing includes tensile and three-point bending tests that induce the part to fail while the load is recorded at which the failure occurs. These values can give indications about the strength of the part and weld seam. The Charpy impact tests are often used in order to measure the impact toughness of weld seams. The crack initiation and development can be recorded, while the cracked surface gives indications about the crack propagation mechanism. Furthermore, peel, pressure, and fracture tests are used. On a large scale, whole components are tested in crash tests. For some applications, it is necessary to evaluate the long-time stability. This is done in fatigue tests, where cyclic loads can be applied.

Besides mechanical destructive tests, chemical tests are also necessary, when the part is used in aggressive environment. Due to the thermal treatment and development of the HAZ in and around the weld seams, the material behavior can deviate from the bulk material's typical reaction on environmental impacts. Therefore, several environmental impact tests using aggressive fluids are developed. Corrosion tests using fresh or salt water are a prominent example. Hydrogen tests can indicate the resistance to hydrogen embrittlement during the use of the material.

For a deeper evaluation of the microstructural development in and around the weld seam, cross-sectional analyses can be performed. These are particularly necessary and used during material and process development to identify the weld seam microstructures and geometrical dimensions, and areas. The preparation includes cutting of the weld seam, grinding, polishing, and often etching in order to make the features visible. For a general analysis of grain sizes and metallurgical characteristics, visual microscopic analyses are used. For weld seam analyses, the hardness is often measured as a relatively simple method to gain information about material properties giving indications about the performance. Depending on the purpose of the evaluation, cross sections can be prepared in both longitudinal and perpendicular directions of the weld seam. Perpendicular cross sections show weld seams at one particular position, while longitudinal cuts show variations during the processing.

Certain etching techniques enable visualization of grain growth directions. This information gives insight into grain development and solidification characteristics.

Scanning Electron Microscopes (SEMs) are particularly useful in order to gain information about detailed surface structures in higher resolution and a higher degree of detail. E.g., weld seam cross sections as well as fractured surfaces can be evaluated. Additionally, energy-dispersive X-ray or wavelength dispersive X-ray spectroscopy (EDX/WDX) methods offer the possibility to derive the local chemical compositions. Often line scans or area scans are used to map the material properties.

Non-destructive testing (NDT) includes methods that gain insight into the material properties, while the welded part can be used for its intended use afterward. NDT methods include ultrasonic NDT, radiography NDT, eddy current NDT, and magnetic particle NDT. Most commonly used for weld seam analysis are probably ultrasonic or X-ray computer tomography (CT) analyses. Ultrasound can be applied by an emitter at a certain location on the part to be tested. At another position located behind the feature to be measured, the receiver collects the spatially resolved ultrasound transferred. Based on the recorded data, conclusions about weld seams can be drawn. Since ultrasound transmits differently in different medium, the existence and locations of voids, pores, or even cracks can be evaluated. X-ray CT can be used to create 2D frames that show density differences in the material. When illuminating the part from different angles, even 3D recreations can be achieved. These 3D frames can show, e.g., inhomogeneities of mixing in the weld seam when material with different densities is used. In addition, porosity is well visible with the additional information of pore dimensions and sizes.

3.7.3.2 Governing Welding Parameters

The process efficiency also defines the economic efficiency. On the one hand, this means that a fast process with low energy and material need is, in general, desired. On the other hand, the weld quality must be achieved, which is defined by the process window of each welding task. The process window is the range of parameter sets usable to achieve the desired welding result. Individual parameters have an impact on the result, but in most cases, the interplay between parameters must be considered that the process is feasible and shows acceptable results. Due to relatively large number of parameters for laser welding processes, the impact of many parameters is needed to be known to find a robust parameter set from the multi-dimensional impact possibilities.

However, often some parameters are pre-defined or restricted from technical reasons, e.g., the kind of gas to be used or the maximum or minimum available laser power or robot movement speed. The most important window for laser processing, in general, and laser welding, in particular, is the relation between the laser power and the processing/welding speed (Figure 3.28). This relation defines the line energy which mainly defines the energy input into the material and thereby the dimensions of the resulting weld seam. For deep penetration welding processes, the lower energy input threshold is the energy needed to reach vaporization and the upper limit is the plasma formation. Another common upper limit for the achievable processing speed is, e.g., the humping that occurs at high welding speeds and can create increased numbers of pores or spatters.

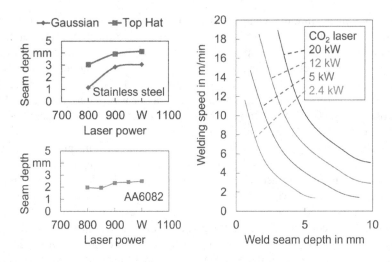

FIGURE 3.28   Visualization of impacts of selected laser welding parameters.

The main welding parameters in laser beam welding are:

- *Laser beam properties*
  - Wavelength: The wavelength is one factor to define the absorption of laser beam energy on the material's surface.
  - Laser beam spot size: The spot size on the material's surface defines the interaction area and thereby the intensity of the laser beam.
  - Divergence: The divergence defines the opening of the laser beam and thereby the sensitivity to misalignment on surface variations.
      The impact of defocusing the laser beam during deep penetration welding is shown in Figure 3.29.
  - Spatial intensity distribution: The intensity distribution within the laser spot defines the energy input distribution into the material.
  - Polarization: Polarization can influence the direction-dependent absorption. However, many laser beams for macro-processing are randomly polarized.

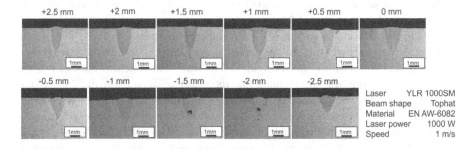

FIGURE 3.29   Cross sections of deep penetration weld seams at varied focal position.

- Inclination angle: In order to avoid back reflections of the reflected laser light back into the fiber, the laser optic is often slightly tilted. Depending on the process sensitivity, the asymmetry must be considered during process design.
- *Laser power*: The laser power defines the laser beam intensity on the surface and the line energy. In a first approximation, the keyhole diameter during laser deep penetration welding is of similar dimension as the beam diameter on the surface.
- *Welding speed*: The welding speed is the relative movement between the laser beam and the work piece and defines the line energy. At low welding speeds, a higher energy input and, in general, wider/deeper weld seams are created.

  Typical weld seam cross-sectional views are shown in Figure 3.30 at varied laser power and welding speed values.
- *Shielding gas*:
  - Type of gas: The kind of shielding gas can influence the protection possibilities of the processing zone against the ambient air and can also alter the process due to chemical reactions with the melt pool or by changing surface tension properties.
  - Gas flow rate: Shielding gas flow rates should not be too weak and not too harsh as well. Too little flow can limit the shielded area and too high flow rates can create turbulences and thereby entrapment of ambient gases.
  - Nozzle arrangement: Coaxial nozzles enable the direction-independent gas application, but require more effort for integration into the setup,

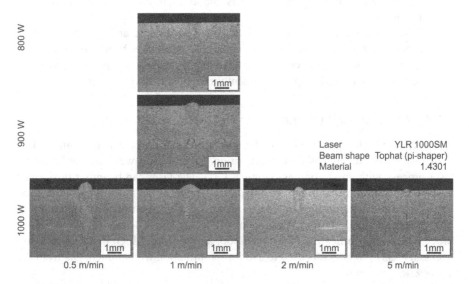

FIGURE 3.30 Cross sections of deep penetration welded seams at varied laser power and welding speed values.

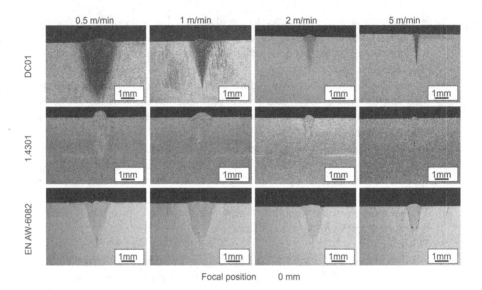

FIGURE 3.31   Cross sections of deep penetration weld seams at different welding speeds and materials.

since the laser beam needs to be guided through its center. Lateral nozzles are more common, while different outlet shapes can be used that must be directed toward the processing zone.

- *Material properties*: Material properties define, among others, the absorption of laser energy and the weld seam geometry.

  A comparison between two steels and an aluminum alloy is shown in Figure 3.31.

- *Joint type and tolerances*: The laser beam positioning and path planning depend on the joint type. Due to comparably small laser beam dimensions, the laser beam and the processing zone require correct and precise clamping for a robust and proper alignment throughout the processing.

3.7.4   LASER-HYBRID WELDING

One variation in the laser welding technology is to combine heat sources and utilize their respective advantages in a combination. Laser-hybrid welding combines the laser process with an arc process (Figure 3.32). The laser can be either positioned leading or trailing to the arc. Typically, a trailing laser beam leads to better mixing of the filler material applied by the arc process but requires more attention to positioning accuracy. Often, it is recommended to use a leading MIG or MAG process to avoid overheating of the melt pool and trailing if high material rates are necessary. Gas protection is used to avoid contamination with ambient gas from the arc nozzle.

Since the laser beam provides the possibility to create a local high energy input into the keyhole, deep weld seams can be achieved. The comparably wider heat input

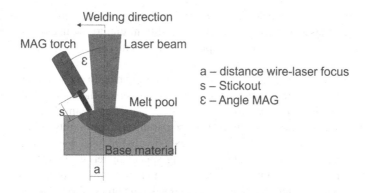

FIGURE 3.32 Sketch of a laser-hybrid welding arrangement.

induced by the arc process decreases the sensitivity to joint tolerances and enables an easier bridging of gaps, but solely to relatively flat heat conduction melt pools (Figure 3.33). Therefore, the combination of both advantageous features is used.

In addition, the two processes have synergetic effects. Since the energy is coupled into the same melt pool, the energy input can be very efficient. At higher arc power, the melt pool temperature is in general higher, which typically increases the absorption of the laser beam and leads to a reduced laser energy needed to achieve the desired weld seam. At the same time, the energy for the phase transformation from solid to liquid (latent heat) is not required to be transferred by the laser beam since the melting has already been achieved by the arc energy input.

From another perspective, the local high temperatures created by the laser beam in the melt pool can increase the process stability. Since the arc searches and burns at the path of lowest resistance from the burner to the specimen, it typically finds varying entry points on the specimen or the melt pool. The induced hot areas by the local laser heating typically show the lowest resistance and provide a stable arc position and process [Möl16].

Besides the mentioned advantages, compromises are necessary. Hybrid laser welding can be conducted at higher process speeds compared to conventional arc

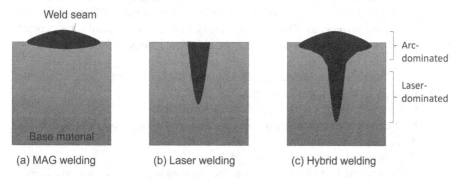

FIGURE 3.33 Schematics of the weld seam appearances at different welding methods.

welding processes. However, the achievable processing speeds are lower compared to comparable laser welding processes. In addition, due to the use of two separate processes, the process is direction-dependent and requires additional effort for alignment and control.

Since two heat sources are active at the same time next to each other, the amount of variable process parameters increases, which influences the melt pool and process characteristics. The laser parameters include laser power, beam diameter, incident angle, polarization, and wavelength. The arc parameters are mainly voltage, current, and stick-out. They must be seen in combination with the typical welding parameters like gas properties and application, welding speed, material properties, and joint design. The increased number of variable parameters makes the process, in general, more complex and an additional effort to find working process windows is necessary. In addition, the use of two systems requires higher investment in the respective systems. Those systems must be usually automated, which can add costs for installation as well. The accessibility of the welding zone can be reduced when adding an arc torch to the process setup.

However, the method enables many process variations. It is possible to enable, e.g., two-sided welding and most joint configurations are possible to perform. Often a better weld quality with reduced spattering can be achieved using hybrid processing compared to single processes. The larger penetration depths and welding speeds compared to arc processes in combination with reduced effort for edge preparation can lead to higher productivity (more melted volume per time compared to single processes), reduced lead times, and less distortion and post-treatment requirements. In addition, the application of filler wire enables the active influence on the weld metal composition and the more efficient gap bridging compared to the sole laser welding process.

3.7.5 Imperfections and Defects

The weld seam quality is of high importance for most welding tasks and must fit the standards applied. Standards define which and how many imperfections at which dimensions are acceptable. There are many imperfections that can occur during and after welding in the weld seam, and also in the surrounding base material due to the effects of heat conduction. Imperfections denote the existence of features that are irregular or inhomogeneous. They can become defects in case they are not acceptable with respect to the required properties. Typical imperfections include (Figure 3.34):

- Cracking (hot or solidification cracking and cold cracking),
- Porosity (process pores or gas pores),
- Inclusions (coatings, dirt, local element accumulations),
- Melt-driven shape changes (undercuts, lack of fusion, drop through, suck back, lack of penetration),
- Deformation (residual stresses), and
- Metallurgical transformations (in the weld seam and the heat-affected zone HAZ).

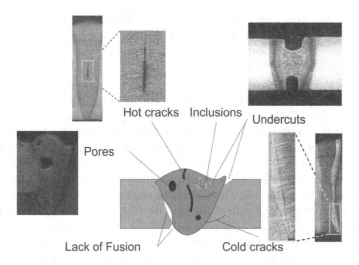

Hot cracks Inclusions Undercuts

Pores

Lack of Fusion Cold cracks

FIGURE 3.34 Typical imperfections and defects in and around a weld seam.

Imperfections are commonly identified after the process either using destructive or non-destructive quality assurance methods. Probably the most common method is the preparation of cross-sectional cuts of weld seams that can be done perpendicular or along the welding direction. After grinding and polishing, microscopic visualization can show the weld seam geometry, porosity, and cracks. A more detailed picture of the surfaces and surface chemical compositions can be achieved by scanning electron microscope (SEM) and X-ray diffraction (XRD), just to name a few possibilities. Non-destructive testing (NDT) can also include micro-CT analysis, while depending on the resolution of the system, even porosity distributions can be visualized.

Nowadays, online process monitoring is gaining more acceptance, which can recognize imperfections during the process. However, the acceptance in industry needs to be gained to trust the online monitoring data to guarantee the seam quality for the customer even without post-process testing. Typically, visual welding cameras are used to identify geometrical variations in the melt pool or to identify spattering. Thermal monitoring is used to recognize changing heat flow conditions that are visible in the surface, which can indicate the occurrence of, e.g., porosity formation or lack of fusion.

Imperfections can have many and often different origins. Some imperfections originate from phenomena occurring in the molten phase of the weld seam. These melt pool and keyhole dynamics can cause, e.g., spatters and initiate pore formation. This makes prediction of their occurrence from the sensor data often difficult. Machine learning, being more advanced, analyzes what indicators in the sensor data can lead to imperfections. In many cases, several sensor data in combination can support the decision-making of the machine learning algorithm to predict and potentially suggest counteractions by adapting the process parameters. Due to partially large amount of data produced by the single sensors, the data streams need to be reduced and analyzed immediately. Sensor fusion techniques that only process and analyze selected features from each sensor data are used. These techniques can support the acceptance of industry when imperfections are robustly and reliably detected and even healed without additional manual effort.

3.7.5.1 Spattering

Spatters are melt ejections originating from the melt pool that can be initiated by melt pool or gas dynamics. Depending on their origin and formation mechanism, spatters can have very small dimensions of nanoscale. They are formed, e.g., by shear forces of the exiting vapor on the keyhole walls and are transferred within the vapor until they slow down and distribute in the ambient atmosphere or condensate and accumulate to larger particles. On the other hand, spatter can show comparably large dimensions of milliscale. These ejections originate from melt pool instabilities due to the induced melt flow patterns that can overcome the restoring forces from the surface tension of melt pool. Melt pool instabilities can also be caused by chemical reactions, e.g., oxidation at insufficient gas protection or surface pollution. Typical causes for spatters can be, e.g., a too high energy input due to a high laser power or low welding speed (Figure 3.35).

Within the keyhole, the melt detachment mechanisms releasing small melt pool drops from the keyhole walls are assumed to originate from the pressure difference between the inside and outside of the keyhole and the induced shear forces due to the high-speed vapor striking the keyhole walls (Figure 3.36).

In a melt pool, surface waves are assumed to be able to build up inertia to overcome surface tension and extract melt pool spatters, e.g., when eigenfrequencies are initiated. Local melt pool accelerations can be initiated also from sudden pressure changes in the keyhole or keyhole movements, e.g., a keyhole collapse or local, temporary, directed vapor flows toward the rear melt pool. Due to the, in general, decreasing surface tension and viscosity at increasing temperatures, the detachment of melt drops becomes easier at higher temperatures. A lower surface tension leads to less counteracting forces to the melt pool inertia. Similarly, low viscosity encourages the melt flow, and less energy is needed to move the melt to form a drop or droplet.

Strategies to avoid spattering mainly aim to calm down the melt pool and avoid dynamic and high-velocity gas flows. This can be achieved with different methods

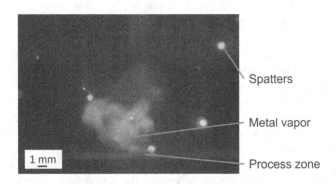

FIGURE 3.35 High-speed image of a welding process.

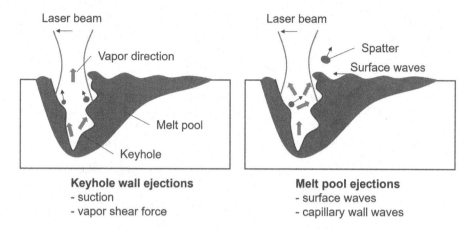

FIGURE 3.36 Sketches of spatter origins and ejection effects.

that provide more space to the melt to flow at lower speeds in the melt pool or give the vapor more space to expand. One possibility is the use of an additional gas flow directed toward the keyhole to widen the keyhole opening. The increased area and often the smoother keyhole opening can reduce the ejection speed of the metal vapor and even prevent the keyhole from collapsing. The smoother vapor outflow can reduce the shear forces and the vapor impact on the keyhole wall. Applied gases can also alter the surface tension of the melt pool surfaces. Higher surface tension can help to limit the melt dynamics. In addition, electromagnetic volume forces can be induced by electromagnets (e.g. [Gat12]) to stabilize the melt pool movements by inducing flow fields against the ejection direction of spatters. A simple mechanical method to suppress spatters is a melt support (e.g., ceramic plates) that are typically positioned on the root side of the welding sheets to form a barrier and can help to avoid root melt ejections.

Another possibility for process stabilization is laser beam shaping, which is largely available for high-power laser processes. Using multiple spot laser beams can induce wider and in turn calmer melt pools that create slower melt flows and thereby avoid ejections, e.g., double focus [Vol16b] or circular spots with additional rings have been shown to provide this effect. Refractive beam shaping promises reliable high-power beam shaping with mirrors or lenses [Las14]. Diffractive optical elements can be used as well [Liu04]. However, the impact of different beam shapes on welding processes is not yet completely understood since multiple effects are induced at the same time. When the energy input is changed, the keyhole shape must adapt, leading to varied energy absorption in depth and temperature distributions that alter the temperature-dependent material properties and the induced melt and vapor flows.

Several investigations have been performed to learn about the relation between the keyhole and melt pool characteristics and the melt ejections, e.g., the spatter amount has been evaluated in high-speed videos using a spatter tracking routine [Vol17]. Refractive beam shaping optics have been used to generate different spatial

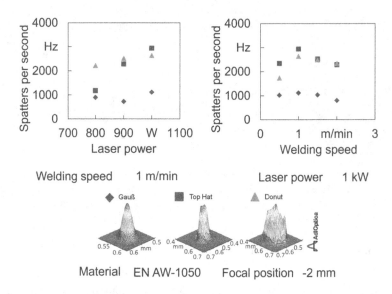

FIGURE 3.37   Spatter occurrence at different laser powers and welding speeds at different beam profiles.

laser intensity distributions on the material's surface, namely top-hat and donut profiles. Figure 3.37 shows the occurrence of spatters at different parameters, while the Gaussian beam shape shows the least number of spatters.

Beam shaping can also be done in beam propagation direction. A bifocal optic has been used to produce two beam waists along the laser optical axis (Figure 3.38a), while the additional high intensity of the second beam waist inside the keyhole can

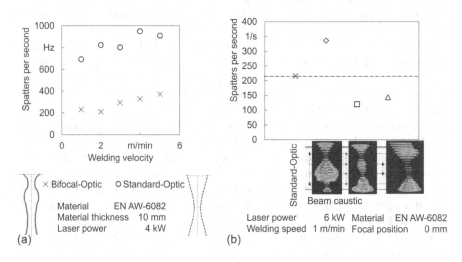

FIGURE 3.38   Spatter occurrence at (a) bifocal beam shaping and (b) multi-focal beam shaping compared to a standard-optic welding process [Vol19a].

stabilize the process [Vol16b]. In addition, a multi-focus optic (foXXus, Adloptica) has been used to produce either a two-waist laser beam with waists close to each other, four waists and two-waists close to each other (Figure 3.38b). Promising stabilization effects have been observed when high-intensity regions are positioned inside the keyhole. It is assumed that the additional energy input increases the vaporization and helps avoiding the collapse of the keyhole that induces intense melt flows and ejections [Vol19a].

Another method to induce a melt pool stabilizing effect has been found when creating a so-called buttonhole. A buttonhole is a melt opening, which is larger compared to a typical keyhole. The buttonhole is solely maintained by surface tension effects instead of the effect of vapor pressure in a keyhole. When the hole expands, it can form curved surfaces showing a catenoid shape. Such a shape is stable due to the surface tension contributions of the two curvatures. It has been shown that beam oscillation patterns [Cho20a] and static beam shaping [Vol21b] can initiate a keyhole elongation and create buttonhole. Calm melt pools and limited impact of vapor have been seen when welding in the buttonhole mode. A buttonhole is self-maintaining and can therefore form a hole in the material when the process is stopped.

3.7.5.2 Humping

The humping effect is a phenomenon where melt pool accumulations occur regularly on the weld seam surface. This phenomenon is typically created when long and flat melt pools are present during the processing. This is typically the case at very high welding velocities (e.g., [Tan20]). The effect is reproducible and is based on the surface tension effect that accumulates melt to humps. The initiated melt flows toward the rear sections of the melt pool support melt accumulation. At high welding speeds, the melt pools can elongate and show very low depth. When a melt section in the rear part of the melt pool is separated from the melt provision from the processing zone, the remaining melt forms a droplet driven by the surface tension. After breakup, the melt pool elongates again until another droplet is formed. This process repeats regularly and forms the droplet appearance in nearly constant distances.

3.7.5.3 Undercuts

Undercuts denote missing material on the weld seam surface (Figure 3.39). They can appear at the top and also on the root side of the weld seam. Possible reasons can be

- melt that is pushed into the joint gap due to thermal expansion of the joining partners,
- the misalignment of the laser beam and the joining gap,
- a too high process velocity relative to the power level and material thickness, or
- surface tension forming a round weld seam shape including material from the weld seam sides.

Undercuts can be accepted to a certain extent though they mark a reduction in weld seam thickness. The lack of material can reduce the weld seam strength compared to the expectations during load application. More importantly, sharp transitions and

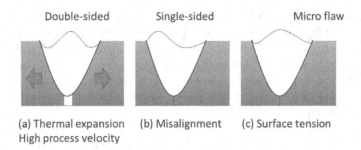

Double-sided   Single-sided   Micro flaw

(a) Thermal expansion   (b) Misalignment   (c) Surface tension
High process velocity

FIGURE 3.39 Schematic sketches of undercuts due to (a) thermal expansion and high process velocities, (b) misalignment, and (c) surface tension.

edges can be a structural notch. These are critical when applying dynamic load and can initiate, e.g., fatigue cracks. In addition, notches can be initiation and starting points of corrosion.

In general, undercuts can be avoided by providing enough material to fill the gaps and form rounded edges. This can be achieved by using enough filler material. When reducing the processing speed, the additional time for the melt to form round melt pools and wet edges can help avoid undercuts. When one-sided undercuts occur, realignment of the laser beam might solve the issue. In addition, magnetic support can help to hold the melt in place and avoid wavy melt pool surfaces. In order to increase spreading and wetting of the melt, surface active elements can be added as part of the filler wire to reduce the melt surface tension.

3.7.5.4   Incomplete Penetration

Incomplete penetration is a lack of material bonding, which reduces the weld seam cross-sectional connection. Some of the most typical joint configurations with incomplete bonding are shown in Figure 3.40. Even though the top view of the weld seam can look acceptable, a lack of material connection can occur underneath. These weld seams are also called "false friends", since they appear to be sufficiently done, but the failure is not immediately visible. This is particularly critical when the lower sides of the weld seam are not visibly accessible or not tested, e.g., by NDT methods.

Possible reasons to initiate incomplete penetration can be changed melt flows that avoid the melt to flow into lower melt pool regions. This can be caused by oxidation effects that change the surface tension and the Marangoni flow. Often, the energy input is too low to reach the required melting for full penetration or to create

Incomplete penetration

FIGURE 3.40 Incomplete penetration defects at different joint configurations.

sufficiently large melt pools. This can be caused by fluctuations of the laser power or a focal shift. Contaminations of the optics or in the shielding gas can be other reasons.

3.7.5.5 Porosity

Pores denote a lack of material and can therefore alter the mechanical properties of welded joints. However, often a certain number or a certain volume of small, spherical pores can be accepted to a certain extent depending on the material, application case, and used standards. Porosity can be categorized in two main categories based on the formation mechanisms, namely metallurgical and process pores.

3.7.5.5.1 Metallurgical Pores

Metallurgical pores are formed by the changing solubility of gases in the melt at different temperatures of the material. Typically, during cooling from the high process peak temperatures, the gas solubility decreases, and gas is extracted from the melt forming gas inside the melt pool. The gas bubbles can accumulate and form pores. Typically, they appear at the liquid-solid interface, where a lower gas pressure is required to start the formation of gas bubbles. Such metallurgical pores are typically comparably small and are often not critical toward global weld seam properties. Typical gases in the melt that induce pore formation during the cooling time are, e.g., nitrogen in austenitic steels or hydrogen in aluminum alloys (Figure 3.41). Sources of such gases are the ambient air including moist, oxides, or other surface layers on the material, and already trapped gas in the previously existing pores.

3.7.5.5.2 Process Pores

Process pores consist of trapped gas from the environment that is included in the melt pool during the processing. Compared to metallurgical pores, the gas that forms

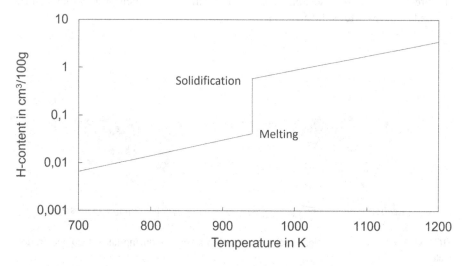

FIGURE 3.41 Example of hydrogen solubility in aluminum.

process pores does not come from the material solution but from the ambient atmosphere. The entrapped gas forms gas bubbles inside the melt pool. In case, the gas cannot escape from the melt pool until solidification, a pore remains in the weld seam. Gas inclusions arise when a gas bubble can form and is surrounded by solidifying material. Spherical pores (<0.2 mm diameter) exist in most fusion weld seams and usually spread homogeneously all over the weld seam. Small, spherical pores are often tolerable according to the standards. Critical pores are long, non-spherical, and large. These characteristics significantly reduce the connecting cross sections of the weld seams. Pores that solidify right on the material's surface can also be seen as critical, since the material response to mechanical load can be affected, e.g., crack initiation can be caused.

Since entrapped gas bubbles in the melt pool have a lower density than the melt, gas bubbles can escape due to buoyancy forces and melt pool flows that can transport the gas bubbles to the surface. In case there is enough time, and the melt flow patterns are not hindering their rise, the gas can escape. At higher density differences between the gas and melt, gas bubbles can rise faster. Since the density ratio is higher for iron compared to, e.g., aluminum, the probability of gas bubbles escaping is higher. In addition, oxide layers can hinder the bubble escaping. In general, larger gas bubbles can rise faster at lower melt viscosity.

Gas entrapment can happen when the melt is able to entrap and surround ambient gas. This can happen during heat conduction welding as well but is more likely to happen in deep penetration welding. In deep penetration welding, the keyhole is present and melt pool dynamics occurs.

Keyhole collapses support the creation of gas bubbles. Due to the pressure fluctuations based on the keyhole dimension oscillations, ambient gas can be sucked into the keyhole for a short time even against the exiting vapor flows. Two basic effects of keyhole collapses are known that can surround the entrapped gas: (1) Keyhole collapse in a middle section and (2) bulging of the rear keyhole and detachment (Figure 3.42). In both cases, gas bubbles are formed in the melt pool, which can be trapped as pores after solidification.

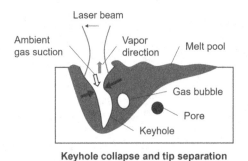

Keyhole collapse and tip separation

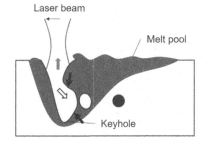

Keyhole bulging and detachment

FIGURE 3.42 Pore formation mechanisms within a keyhole during laser deep penetration welding.

Further aspects that can impact the pore formation need to be considered. The type of shielding and ambient gas define the probability and severeness of porosity, e.g., due to oxidation effects. Furthermore, surface contaminations, alloying elements, or coatings can change the process dynamics and can even provide gas to be entrapped. Zinc coatings can affect the process. Zinc has a lower vaporization point as the steel base material and forms additional gas pressure in the melt pool that can lead to increased dynamics, pore formation, and also spatter ejections. The produced gas is an additional source for gas being entrapped.

Also, the applied shielding gas flow can impact gas entrapment mechanisms. Effects and material properties that increase the process dynamics typically support creation of pores. A low viscosity of the material at high temperatures (e.g., aluminum alloys) or the vaporization of alloying elements increases temporal fluctuations of process dynamics, increasing the probability of a keyhole collapse. Porosity is most commonly found in aluminum or nickel alloys. Also, steels with medium and high C-content are affected.

Prevention mechanisms are based on

1. enabling gas bubble escaping by
   a. creating long melt pools to encourage the gas bubble movement to the melt pool surface,
   b. a second treatment for re-melting (e.g., another re-melting with the laser beam).
2. decreasing keyhole collapses by
   a. laser power modulations,
   b. filler wire use,
   c. a gas flow directed toward the rear keyhole wall, and
   d. magnetic stabilization.
3. eliminating gas availability (e.g., by welding in low-pressure or vacuum conditions).

3.7.5.6 Cracking

Cracks are material separations that occur during (hot cracking or solidification cracking) or after the process (cold cracking). Their occurrence can reach from micro to macro dimensions and are typically not tolerated from a quality perspective due to their negative impact on the properties of a material. Cracks can occur at different positions within the weld seam and heat-affected zones. They can occur as inter- (through grains) and trans-crystalline (along grain boundaries) cracks.

3.7.5.6.1 Cold Cracking

Cold cracks occur after solidification of a material when local stress fields separate the material. They form due to three main factors: (1) Presence of hydrogen in the material, (2) a brittle microstructure, e.g., martensite, and (3) stress concentration. This can happen right during the cooling phase but can also occur days after welding. Cold cracks can appear in the weld seam and the HAZ. Due to the formation mechanism, the surfaces of the cracked material parts appear to be rugged.

Cold cracking can often be reduced and avoided by

- pre-heating the material in order to reduce thermal gradients in the material, thereby decreasing the creation of high stress concentrations,
- controlled cooling conditions (e.g., slow cooling by controlled temperature decrease),
- filler material addition to alter the chemical composition for grain refinement or viscosity reduction (e.g., TiB_2 in aluminum),
- widening the temperature field by increasing the line energy to lower the cooling rates and reduce the development of thermal stresses, or
- multi-layer welding to induce intrinsic pre- and post-heating.

3.7.5.6.2 Hot Cracking

Hot cracking appears during melt solidification in the mushy zone, where thermal stresses can lead to a separation of partly solidified material. Hot cracks typically show a smooth, round surface since the separation happens in partly liquid state and final solidification happens thereafter.

The mushy (partly melted) zone occurs due to constitutional undercooling induced by chemical elements in the alloy. For pure materials, there is one single melting point, while for alloys, solidification can happen within a temperature range. The alloy elements can increase the range between the liquidus and solidus temperatures. Therefore, a mushy zone can form, where parts of the mush is solid and some parts are still liquid. This means that grains start growing and form the lattice during solidification. At the same time, the low melting elements accumulate in the remaining liquid. This leads to the effect of constitutional undercooling, continuously increasing the solidus temperature in the remaining melt. In addition, this leads to an inhomogeneous distribution of the elements in the weld seam after solidification. Therefore, alloying elements are mainly found on the grain boundaries. Due to thermal shrinking of grains from liquid to solid state, cavities form between the solidifying grains. These cavities create low-pressure volumes that the liquid tries to fill by back feeding (Figure 3.43). However, the permeability of the already existing solid grain structure can be low and it avoids the back feeding to fill the cavities. Small grains support permeability, while larger grains tend to block back feeding. In case the loads, e.g., from clamping or thermal stresses due to shrinking are high enough, the partly

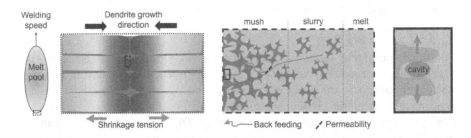

FIGURE 3.43 Schematics of the grain growth in the mushy zone.

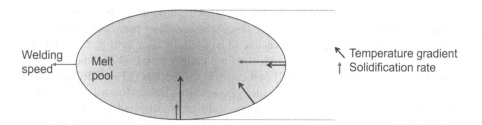

FIGURE 3.44 Temperature gradient and solidification rate at an exemplary melt pool in top view.

solidified material can separate. Since the material is still partly liquid, the crack surfaces appear rounded due to surface tension.

The mushy zone dimensions and grain growth characteristics depend on the material alloy and the local temperature gradients and solidification rates (Figure 3.44). Small grains can be initiated when grain initiators are present in the alloy, when the temperature gradient is high and the solidification rate is low.

When the difference between the liquidus and solidus temperatures and the viscosity are high, cracks are more likely to occur. Hot cracks can occur as central weld seam cracks and as macro and micro cracks between dendrites.

Carbon, sulfur, and phosphorous as alloying elements in steel increase the risk of hot cracking, while manganese decreases the risk. Due to the typically larger range between solidus and liquidus temperatures in aluminum alloys, there is a high risk of cracking. High amounts of silicon and magnesium increase the crack susceptibility due to increased amounts of low melting phases.

Hot cracking can be reduced or avoided by

- changing the metallurgy (e.g., addition of grain refiners like titanium in the alloy),
- changing the geometry of the structure to avoid stresses, or
- thermal design by lowering the cooling rate and shrinking tensions (e.g., by pre-heating the specimen or local heat sources) or energy input design by, e.g., beam shaping (e.g., beam oscillation to create a wider melt pool).

3.8 LASER BEAM CUTTING

3.8.1 INTRODUCTION

Laser cutting has been already invented at TWI, Cambridge (UK) in 1967 and is categorized into the separation processes. Laser cutting is still the most industrially used high-power laser technology since it is comparably easy to use and can be used for many materials. High process velocities and good cut quality can be achieved. Due to low-energy input compared to competing cutting processes, a small heat-affected zone can be achieved. The laser cutting heads are typically mounted on robot or CNC-machines, while cutting of sheets is the most common application for various industries. Both CO_2 and YAG lasers are possible to be used. Conductive sensors are often installed to measure and control the distance of the gas nozzle and the material to be cut. All materials that absorb the used laser beam and can establish a local melt pool are, in general, feasible to be used for laser cutting. Materials for laser beam cutting in industrial production are mainly metals (40%) and polymers (30%), followed by ceramics (18%) and ceramic composites (12%). The ejection of the melt is achieved by an impulse using gas pressure directed toward the melt pool or the laser-induced recoil pressure. In gas cutting, the directed gas outflow from a gas nozzle is used to create pressure on the melt pool, which is positioned close above the material. Inert gases and oxygen are typically used as cutting gases.

3.8.2 PROCESS PRINCIPLE

The principle of laser cutting is based on the local melting of a material and exhaust of the molten material by either a directed gas jet or laser ablation (Figure 3.45). Typically, high intensity laser beams, often in focal position, are used to generate melt pools. The high intensity of laser beams enables the creation of deep and narrow impacts, similar to deep penetration welding processes. This mode can be achieved when overcoming the deep penetration threshold of the material. However, during cutting, the molten volume does not remain as a trailing melt pool as it is intended for laser beam welding. The typical cut front is formed that is illuminated by the laser beam. Similar to keyhole welding, during laser cutting, a downward movement of the melt on the vaporizing front is observed. This melt flow can support the ejection of melt when gas supported cutting is applied or can be the main driver and force to expel molten droplets from the cut front root side.

The cut front allows a high-energy input and the creation of a deep melt pool to achieve the cutting of even thick sheets. The energy input must balance the main heat loss due to heat conduction, while the process velocity enables the movement of the laser beam relative to the material and a continuous exhaust of the material.

The energy input into the material takes place due to Fresnel absorption on the created cut front and side walls. Multiple reflections can occur and increase

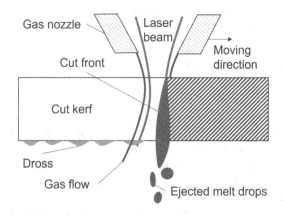

FIGURE 3.45 Principle sketch of the laser cutting process.

the total absorbed energy depending on the keyhole shape. When the cut front is shaped to enable the laser beam rays to reflect toward a side wall of the molten pool, multiple reflections and absorption can support the energy input. However, the multiple reflection and absorption effect is less efficient compared to laser deep penetration welding since there is no rear wall that can absorb the reflected rays. The created melt film governs the process behavior. The melt film thickness is defined by the absorption of the laser beam, which defines the downstream speed. The cut front inclination angle mainly depends on the processing speed; however, the angle is steeper at slower process velocities. Due to the steep inclination of the cut front, parts of the laser beam can miss the material, which decreases the process efficiency.

In addition, the selection of gases as a cutting gas to remove the molten material can change the energy balance. Passive gases, which are typically inert or at least low-reactive gases, do not chemically react with the melt pool components. These gases can have a minor cooling effect. However, when using active gases, such as oxygen, the induced chemical reactions induce additional energy to melt the material. This reduces the laser energy needed to create the necessary melt pool dimensions, which can be more cost-efficient due to typically lower price for O_2 compared to the cost of laser energy. The impact of the reaction can, however, increase the process dynamics and the risk of burning. Using oxygen can support the formation of oxides that can remain on the cut kerf and alter the surface appearance and kerf properties.

The overall energy input must be sufficiently high to heat the material and overcome the latent heat of melting. In sublimation cutting, in addition, the melt must be further heated, and the latent heat of vaporization needs to be overcome, which is typically ten times higher than the one for melting [Pop11].

When the melt pool was formed due to the energy input, the created melt film at the cut front is then forced by a gas jet or laser recoil pressure to downstream. The induced melt flows typically lead to ripples on the cut kerf. The roughness of the solidified cut kerf is one of the quality criteria of laser cutting results.

On the root side of the cut front, melt is ejected. Depending on the flow rate and material properties, the separation of the melt from the melt pool happens from a liquid film or direct droplet formation and detachment. The downstream inertia in the melt pool must overcome the local surface tension to create a detachment into droplets.

Another important quality factor in laser cutting is the formation of dross. Dross is the material that is transferred in the cut front but solidified on the lower side of the sheet. This happens when the induced gas or recoil pressure guiding the melt is not sufficiently directed or not strong enough to detach the material. Dross can form if the molten drop's inertia is not sufficient to overcome the surface tension.

The introduced heat from the laser beam into the process zone is balanced by energy losses occurring due to heat transfer via conduction into the base material as well as via convection within the melt pool. Convective melt pool flows are mainly induced by laser ablation on the cut front. Melting and evaporation occur, including re-condensation and re-solidification, after the laser beam is passed. Chemical reactions and element vaporization can lead to changes in the chemical composition of the material. In combination with the thermal cycles induced, microstructural changes need to be considered in the vicinity of the cut kerf.

3.8.3 PROCESS MODES

Three main process modes can be distinguished, namely inert gas cutting, oxygen cutting, and vaporization cutting:

- In laser inert gas cutting, typically nitrogen and argon at pressure in the range of 2–20 bar are used to induce material ejection. The process can lead to shiny edges and no burning damage but is comparably slow and a high laser energy is necessary.
- Laser oxygen cutting uses the energy created by surface oxidation as additional energy input. Therefore, a reduced laser energy is possible to be used compared to inert gas cutting. The cut kerf must be expected to be oxidized after the cutting process. The process velocity is comparably high or high thickness material can be cut. High pressure in the range of 6 bar is used to remove the melt. Cutting of material of 40 mm thickness is possible.
- In laser vaporization cutting, no cutting gas is used to remove the material. The laser-induced recoil pressure on the boiling cut front directly induces the melt downward flow and ejection. This setup enables remote cutting, where oscillating optics can control the laser beam position rapidly without the limitation of moving the whole optic including the gas nozzle. Remote cutting is typically used to cut metals as well as non-metals (e.g., plastic and wood).

One process variation is laser waterjet cutting, where water is used for melt removal. In addition, the laser beam can be efficiently guided inside the water stream comparable to the light transfer inside a glass fiber. The water supports the cooling of the material, which can reduce the cracking and can avoid the attachment of ejected particles. Very good cut quality can be achieved but the process velocities are comparably slow.

3.8.4 PROCESS PARAMETERS

Typically, high-power laser systems (CO_2 and YAG lasers) are used for laser beam cutting in cw mode. For micro-cutting applications, the pw modes are also used. Both bell-shaped and top-hat beam shapes can be applied; beam shaping can help to direct the energy input on the cut front and increase the homogeneity of the melt pool downward movement.

Laser beam polarization can have a significant impact on the cutting result due to varying absorption when cutting in different processing directions. Linear polarization can lead to varying cut quality and cut appearances. Linear p-polarized beams in moving direction have the highest absorption on the cut front and when turning the moving direction by 90°, absorption happens mainly on the cut front side walls. Therefore, circular polarization or multi-mode beams with no preferred polarization are used to avoid direction-dependent effects. Typical material thicknesses

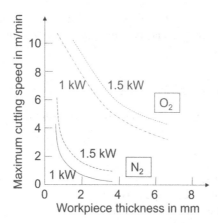

FIGURE 3.46 Example of the tendencies of process parameter impact on maximum cutting speed (mild steel).

range between 0.1 mm and 20 mm. Cutting parameters range from below 1 m/min to 100 m/min processing velocity and 1 kW to 8 kW laser power. Typical tendencies of parameter impact on the cutting speed are shown in Figure 3.46.

The pressure of either the assist gas or the recoil pressure from the laser beam changes the melt pool geometry and melt film movement. In general, a higher pressure leads to a fast melt film downward movement and a steep melt pool front angle (Figure 3.47).

The laser wavelength influences the process as well. At high wavelengths (e.g., from CO_2 laser systems), the laser beam absorption is typically low. However, these laser systems can reach high power levels and are reliable. High-quality results can also be achieved with lasers at 1 μm wavelength. Many systems installed in industry use the reliable CO_2 lasers, but fiber and disk lasers become more popular in particular because of the possibility to guide the laser beam in fiber optics to the processing zone. In addition, the absorption is, in general, higher for most materials using YAG lasers compared to CO_2 lasers.

In addition, laser beam caustic parameters influence the process result. When using a laser beam with small beam diameter on the material's surface, a higher intensity can be reached, which enables, in general, higher process velocities. The

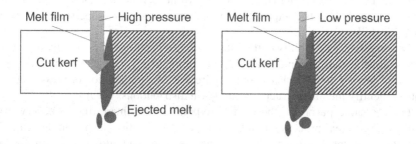

FIGURE 3.47 Melt film appearance at different gas or recoil pressures.

positioning of the beam waist slightly below the material's surface enables the highest possible process velocities and best quality due to the increased absorption at lower parts of the cut front.

3.8.5 QUALITY CRITERIA AND IMPERFECTIONS

Typical quality criteria applied to laser cutting results are geometrical characteristics like

- width of the cut kerf,
- surface roughness,
- rectangularity of the cut edge corners,
- parallel sides of cut surface walls, and
- sharpness of the corners.

Imperfection criteria are

- low dross attachment at the bottom edge,
- low oxidation of the cut surfaces,
- small HAZs,
- no cracks at the cut surface, and
- low loss of alloying elements by vaporization.

The dominant dynamical features of the process are ripple formation on the cut kerf [Fri00] and adherent dross on the root side of the sheets ([Mak92, Nem97]). At least three different ripple patterns exist, namely ripples of first, second, and third kind. Ripples of the first kind appear at the top of the cut edge, while nearly no re-solidified material evolves near the top of the cut edge. The ripple frequency linearly increases with cutting speed up to a value of about 500 Hz at 4 m/min. Second kind ripples show the doubled ripple frequency compared to the first kind and appear in the middle section of the cut kerf. When increasing the cutting speed, they appear closer to the top surface (Figure 3.48). Ripples of the third kind show re-solidified molten material.

Dross is solidified material that is not blown out of the processing zone but attached on the lower part of the cut sheets. Reasons for dross formation can be the dynamic movement of melt in the form of cut front oscillations ([Dew21, Poc17]), too high pressure gradients compared to shear forces [Vic87], or supersonic gas movement [Sch87].

Due to direct accessibility of the cut surface after cutting, a first quality inspection can be done visually. Ripples and dross are immediately visible. Thermal effects in the HAZ are not directly visible and require, e.g., the preparation of cross-sectional cuts to identify the microstructures and grain distributions.

For academic purposes, the melt flow patterns are interesting to observe and explain at different process parameters. Process simulations can help describe the melt flow patterns. Cutting behind glass has been shown to efficiently visualize the melt flows [Dew21]. Some limitations due to different material properties of glass

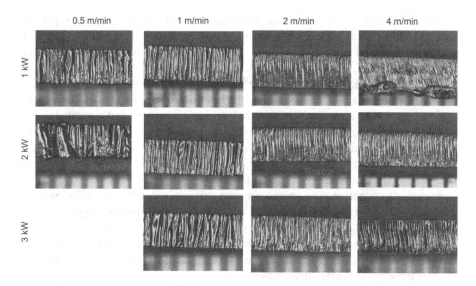

FIGURE 3.48 Cut kerfs with ripples, dross, and burning at different laser powers and processing speeds.

compared to other material need to be considered that alter material properties such as heat conductivity.

3.9 LASER DRILLING

Laser beams can be used for drilling operations. Laser drilling uses high-intensity laser beams in continuous and pulsed modes. In order to achieve the removal of material, local melting and/or vaporization is induced by the laser beam. Drilling methods with conventional mechanical tools with defined shape form the drill holes based on the geometry of the tool. The laser beam has no mechanical geometry. Therefore, the geometry of the drill hole depends on other factors related to the laser beam parameters and processing characteristics in correlation with the laser beam-matter interaction. The laser drill holes are in the dimension of the laser beam diameter, but additional effect from melting and vaporization or beam movement can shape the drill hole geometry. Therefore, on the one hand, the process becomes complex compared to drilling with conventional tools. However, on the other hand, the laser beam parameters open new possibilities, such as drill holes with opening cone geometries, which are very difficult to achieve with conventional methods.

For creating drill holes, local laser melting or vaporization is induced. During laser drilling, the material is always melted to a certain extent, even when mainly the vaporization mechanism is aimed for. In continuous laser operation or long pulse applications, melting is the dominating mechanism. An upward melt movement is induced by the laser recoil pressure in the drill hole, and the melt is ejected upward. Since the melt during solidification reshapes, e.g., based on the surface tension effects, the final drill hole shape can show unwanted deviations from nicely shaped straight walls.

Therefore, high-intense short pulse laser modes are often used to induce vaporization of the material as the main material removal mechanism. This mode reduces the
amount of melt and still removes material as vapor. The vaporized material volume
forms the drill hole. The melt vaporization or ejection is initiated by the laser beam.
There is no additional gas jet used as in gas-assisted laser cutting for melt removal.
Instead, the ablation pressure created by the laser beam performs the ejection of the
material, which is either directly vaporized or melt accelerated and separated.

During laser illumination, the photonic energy is mainly transferred from the
photons to the electrons within nanoseconds to picoseconds after the impact of the
laser photon. This is followed by the energy transfer to the lattice material. When
using short, high-intense laser pulses, the energy can be directly transferred to a
large extent to vaporize the material. In laser drilling, the direct vaporization is an
advantage since the material is removed with minimal melt creation. In principle, in
the vaporization mode, there are more exiting atoms leaving the bulk material than
can re-condensed, which denotes the material removal. The material vapor removal
is, on the one hand, the goal of the process, but is also a source of energy loss and
reduces thereby the surface temperature.

However, even when aiming to vaporize the material using short pulses, classical
heat conduction happens in the drill hole as well, which creates a thermal gradient
from the drill hole surface to the surrounding material. This induces the creation of
a thin melt pool (Figure 3.49a). Extensive melt appearance is one of the main reasons
for imperfections and inaccuracies of the resulting drill hole geometry. The melt film
thickness varies with illumination times. At very short pulses (<100 ps), it is possible
to almost avoid melting and just remove material by vaporization (Figure 3.49b).
These two drilling modes are often used as theoretical models to simulate and predict material removal.

The pulse length of the pulsed lasers and the total absorbed energy define the surface temperature of the drill hole during its formation. The optimum pulse lengths
are material-dependent, e.g., ~1–10 ps (Fe) and ~5 ps (Al). In general, most metals
and some ceramics and plastics can be laser drilled.

The drilling speed is defined as the material removal per pulse in depth direction.
The intensity on the more or less flat drill hole ground mainly defines the removal

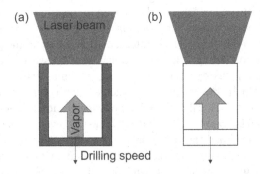

FIGURE 3.49 Theoretical models of laser drilling (a) melting and vaporization and (b) sole
vaporization.

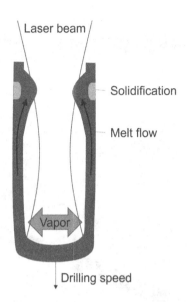

FIGURE 3.50   Schematic sketch of melt pool movement during drill hole closure.

rate. In general, at high intensities, more vapor is created and a higher ablation pressure results, which can lead to a smaller melt film. However, small melt films can be accelerated by the recoil pressure impact and can move at speeds of several 10 m/s [All87]. Such dynamic melt movements can be the reason for process imperfections. There is a possibility that the expanding lower part of the drill hole and the upward melt movement toward the solidifying upper drill hole part can even lead to melt pool closure (Figure 3.50).

Dross imperfections can occur when melt remains at the drill hole exit or on the walls after the laser pulse. Such drill hole shape changes can also induce craters at the lower part of the drill hole or roundness and concentricity changes and variations in depth. In addition, metallurgical changes and cracks can appear. Although heat input into the surrounding material is limited, a HAZ still occurs, which means that the microstructure can be altered within and beside the melt film. Cracks can be initiated due to the thermal impact and related stress fields induced. In particular, the high thermal gradients are a potential source of thermal stresses. Another factor that can alter the energy input and thereby the dynamics of the process is the absorption of laser energy in plasma or ionized gas. Plasma or ionized gas occur at high laser energy input, which can be provided during the high intensity of the short laser pulses. Therefore, such absorption mechanisms can limit the pulse peak intensity and the repetition rates that can be used. In addition, such inhomogeneous and potentially scattered energy input can deform the drill hole and reduce the drill depth per pulse. When using polarized laser beams, influences of angle-dependent variations need to be considered.

Laser drill holes can be produced in different ways. A single pulse can produce a drill hole with only one step, while percussion drilling methods remove material

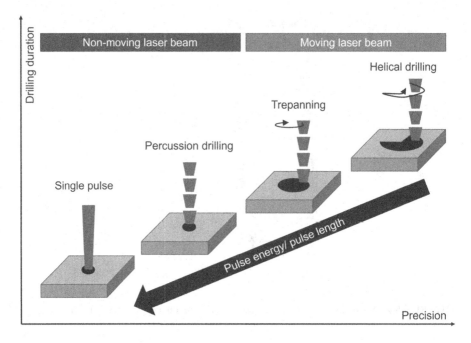

FIGURE 3.51   Categorization of drilling modes.

stepwise in depth during multiple laser pulses (Figure 3.51). More complex laser beam movements can be used for trepanning and helical drilling to remove material stepwise and shape the drill hole at the same time. In order to limit the creation of imperfections, in general, shorter pulses at high intensity are advantageous, which also reduce the energy needed per pulse.

Single pulses are usually used in the range of 10 μs–1 ms pulse duration, reaching drilling depths between 0.02 mm and several millimeters. A single pulse process can manage to create around 1000 holes per second, which shows, in general, a high productivity at comparably low quality.

Percussion drilling uses laser pulse lengths in the dimension of ~10 ns at a laser energy input of 10 J–0.1 mJ. Compared to single pulse drilling, percussion drilling can achieve deeper drill holes at a more homogeneous diameter. The pulsing controls and limits the creation of melt and resulting burr to some extent. However, in this mode, the melt creation and flow influence the process result. The removal rate depends on the material depth due to varying removal rates along the drill hole creation in depth.

Trepanning and helical drilling use a pulsed laser beam, which moves relative to the substrate using beam oscillation optics. That way, in many removal paths, the drill shape is created with many single pulses. The advantage of minimum melt creation per pulse is combined with the flexibility to create nearly arbitrary drill hole shapes. Also, angled beams for varying diameters of the drill hole can be used. Trepanning is typically started with a statically created drill hole, while helical drilling is immediately started with the moving beam. High depth and quality are possible to reach due to the possibility of adapting the focal position during drilling. The adaption to the actual drill depth enables the processing with the intended laser beam spot size in every depth.

The resulting drill quality is usually defined by the ratio between melt and vapor material ejection. The more melt is created, the less precise and less costly the drilling process is. Less melt can be achieved at shorter pulses, requiring however higher beam intensities.

Laser drilling is a non-contact, fast process. No direct wear of tools during the process is created. The use of remote laser beams offers a good accessibility to guide the laser beam to the surface of the part or component that needs to be processed. The process can be fully automated, which enables drilling of a huge number of holes in a short time. Laser drilling reaches large depth to width ratio (>1) and covers a large range of achievable drill hole diameters with the same laser system. Holes at small angles can be drilled with no burr at high accuracy. Many materials can be laser-drilled, e.g., hard diamond, ceramics, or hardened steels.

3.10   LASER ALLOYING AND DISPERSING

The category "laser surface treatment" contains several processes (Figure 3.52) that use different effects to alter material's surfaces. Hardening (see Section 3.3) uses comparably low laser energy input to avoid melting and focus on inducing hardening effects, while other surface treatment processes require melting. The re-melting process of the surface can be applied, inducing a melt pool on the surface. This process can be used, e.g., to decrease porosity of a previously produced weld seam, to enhance material mixing, or to create a certain surface structure. When applying additional material to the created melt pool, laser alloying, dispersing, or cladding can be realized. The processes are very similar since they all contain the addition of material. The differences are the kind of material applied and the purpose of the material addition.

Alloying, in general, follows the purpose of adding additional alloying elements to the base material and thereby altering the chemical composition. That way, it is possible to balance element vaporization that occurs during the surface treatment or change material properties to achieve, e.g., better wear resistance of the surface. In addition, metallurgical changes can be induced that help to avoid imperfections like cracking by refining the grain structure.

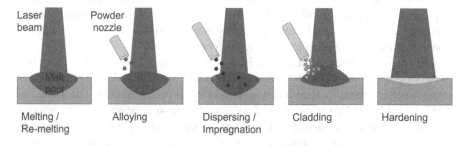

FIGURE 3.52   Process variants of laser surface treatments.

In dispersing, the added material in the liquid metal melt pool is aimed to stay in solid phase. Typically, carbides, borides, or other high-melting materials are added for surface wear resistance or hardness improvement. The additional material can be added using powder nozzles, while a gas flow directs the powder particles onto the melt pool. Also, a two-step process, where the powder is pre-placed on the material before melting is possible to use to incorporate the powder material. The added material is aimed to distribute homogeneously in the surface area to provide the intended surface properties in the whole depth and width of the treated surface volume. In principle, gases can also be used for alloying, e.g., nitrogen reacts with alloying elements in the melt pool that create high-melting phases.

For the mentioned surface treatment methods that include the creation of melt pools, typically, melt pools in heat conduction mode are intended to be produced. Parameters are usually in the range of 1–3 kW laser power, 0.1–10 m/min process velocity, 1–5 mm laser spot diameter on the surface and 5–30 g/min applied material flow rate. Also, the initiation of the deep penetration mode is possible, leading to the process called "laser deep alloying" [Par07]. This process mode can enable a more efficient energy absorptance, wider and deeper tracks, and more rectangular cross-sectional appearance of the tracks. The challenge arises to enable the added material to distribute within the comparably large melt pool during the short duration of its existence. The powder particles are distributed by the melt pool convection into deeper material sections. Beam shaping techniques, such as beam oscillation, have been shown to support the melt mixing and distribution [Vol14].

Large surfaces are treated using multiple tracks next to each other. The hatch distance (distance between two tracks) must be chosen to re-melt the desired volume and to keep the overlap of tracks to a minimum for higher process efficiency. Quality criteria applied to evaluate the surfaces are typically the homogeneity of the distribution of the elements/particles in the melted surface volume by convection and diffusion and mechanical or chemical properties, e.g., wear resistance or corrosion resistance.

3.11   LASER CLADDING

Laser cladding is a technique within the surface treatment methods (Figure 3.46). The purpose of the process is to locally add material to create a surface layer with certain properties. This can be required to change surface properties by adding a layer of material or for locally adding material for repairing of worn surfaces. Typical surface properties that are intended to be influenced are the hardness, wear, and corrosion resistance. Therefore, cladding is typically done as one-layer application consisting of multiple tracks next to each other.

The process is based on material addition onto the base material. However, parts of the base structure or already-built structure below are melted during processing in order to achieve metallurgical bonding. This effect is called dilution. The additional material can be applied in the form of wire or powder. Wire application is comparably easy to control. The amount of material incorporated, and the related chemical composition of the melt mixture is predictable. However, the wire feeding system can make the process direction-dependent, if the wire is not applied coaxially. Powder can be pre-placed or more typically applied by deposition systems. The powder or wire is fed into the melt pool and it liquefies.

Often, surface cladding methods do not require surface preparation before processing. Comparably large laser beam spots on the material's surfaces are used (1–5 mm diameter). Such spots can be either produced by using suitable laser beam guiding and forming optics or by using the beam in defocused position. Moreover, beam shaping techniques are used to increase the treated surface area per time by increasing the molten track width or to avoid local overheating by distributing energy within the spatial intensity distribution to decrease the risk of local overheating, or to create a more homogeneous heat input into the materials. Furthermore, process velocity, material parameters (powder flow, base material properties), and the hatch distance between the tracks define the melt pool dimensions and solidification behavior. The hatch distance between the tracks is typically optimum around 50% of the track width. It mainly defines the surface waviness of the resulting clad layer. In case the hatch distance is too close, bulging can occur, leading to large material accumulations and clad layer height variations. At too large distances between the tracks, the material might not fill the gaps sufficiently and height variations and even gaps can remain on the surface.

Typical track heights are in the range of 0.05–2 mm per layer at a single-track width of 0.1–45 mm using laser powers used in the range of 0.5–6 kW at process velocities in the range of 0.1–10 m/min (Figure 3.53). In order to achieve the desired surface geometry, post-treatment can be necessary, such as milling.

For powder application, powder feeding systems are used. The systems consist of a powder feeder, tubes, and the nozzle at the processing zone. A powder feeder has a powder hopper that can be filled with the powder material and spreads the powder onto a rotating disk. From those disks, the powder particles are taken and guided into the transferring tubes by inert or low-reaction gases to the nozzle at the processing head. The tubes and the powder nozzle define the maximum possible gas and powder flow.

Powder nozzles can be positioned externally from the side, which is an easy way to fix and align the particle flow in relation to the melt pool. The use of coaxial powder feeding systems are more typical. Thereby, the laser beam is transferred through the center of the nozzle to enable direction-independent and flexible processing. The powder speed and divergence of the powder flow define the powder capture efficiency and incorporation behavior of the particles into the melt pool. The melt pool size on the material's surface should be larger compared to the particle flow dimensions to enable the efficient incorporation of powder particles within

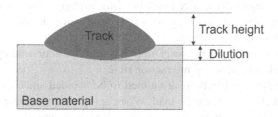

FIGURE 3.53 Geometry of a cladding track.

the whole powder stream. Often, additional shielding gas is used for shielding the process zone from the ambient atmosphere. Modern nozzle systems have this option already included into the nozzle system. Well-feedable powder sizes in common systems are in the range of 20–150 μm. Smaller particles can agglomerate and clog the tubes and the nozzle. Larger particles are often not foreseen to be guided through the nozzle system with its fine internal channels.

From a practical perspective, direct back reflections of the laser beam from shiny flat surface into the beam guiding system should be avoided. Back reflections can easily occur and go back straight along the laser beam axis, and in worst case, melt the fiber ends or even damage the laser machine. Angling of optics, e.g., as done in welding operations would decrease the risk of back reflections but can lead to single-sided effects during the processing due to the related angled powder application. When powder is applied, the laser light is typically absorbed and scattered enough to avoid back reflections. Therefore, it is recommended to start the process only when the powder is flowing to absorb and scatter some of the laser light.

General trends during the cladding process are:

- The more powder is applied, the higher is the front inclination angle of the melt pool.
- Melt pool flows can be directly influenced by the shielding and transport gas due to the induced flow momentum or indirectly by altering the surface temperature and surface tension gradients and thereby the Marangoni flow.
- At too high powder flow rates, the laser beam can be absorbed and scattered by the particles [Gas93]. Limited melt pool sizes and the resulting lack of fusion can occur due to decreased energy input to the melt pool.
- At low powder application (e.g., when the powder flow is interrupted or reduced), the laser energy transfer to the base material's surface can be higher than expected. This can induce back reflections from the flat reflective surface.
- Cracks can occur between the substrate and the track during cooling. Stress release by thermal cycle control can reduce thermal gradients.

Typical quality criteria of cladded layers are the resulting track geometries, imperfections, bonding strength, wear and corrosion resistance properties, and the amount of dilution. Dilution denotes the amount of melted base material required to achieve the bonding of material to the clad track. There are also clad tracks attached on top of a surface similar to brazing processes without melting the base material's surface with the disadvantage of typically lower bonding strengths. Dilution is aimed to be minimum for reaching a high-energy efficiency and minimum contamination of the added track with base material contaminants that can change the chemical composition and thereby material properties. Dilution is mainly influenced by the properties of a material and laser-material interaction time.

Porosity and notch formations are aimed to be avoided since they can be starting points of cracking at dynamic load. When applying multi-materials or dissimilar materials, brittle intermetallic compounds can be created at the transition zone, which can decrease the bonding properties.

With improved laser characteristics (e.g., higher power, better focusability, and beam shaping) and equipment development (e.g., powder feeding systems), cladding in one layer could be expanded to build 3D structures as one technology within AM. Complex multi-layer structures have been enabled, which were first used for rapid prototyping and later in regular production.

3.12   LASER BEAM ADDITIVE MANUFACTURING OF METALS

Traditionally, manufacturing methods to form the final geometry are mainly based on subtractive methods. These include milling, drilling, turning, cutting, or sawing. Thereby, material is removed from a bulk material to derive the final shape of the part. Larger components are joined combining single subtractive parts. Complexity of parts is often limited due to accessibility challenges of subtractive tools. Formative processes like forging, forming, or injection molding are rapid methods to create a final part shape and are practical for mass production, where the same shape is needed to be produced many times. However, there is a very limited possibility to customize parts or create complex shapes.

AM is a generative approach to iteratively create the part layer by layer or section by section. Thereby, material is only applied where needed. This can reduce the material waste and cost for material purchase. Instead of buying a large block of high-value material, it is possible to only purchase the material needed for actually building the part. Parts and components are produced in a generative way with "access" to all material steps and the opportunity to include internal structures and functional features during the production steps. Such an addition of material as process principle can be achieved in many ways using liquid, wire, or powder material with different bonding methods. The principal idea of AM is the creation of the final part from feedstock material. Different methods have been developed that are able to produce parts fulfilling the three Fs: The "form" needs to be correct, followed by the "fit" in relation to other parts in a component, and the "function" must be fulfilled. Most current methods cannot fulfill these aspects completely, and often post-processing is necessary to reach the desired material properties, shapes, and functionalities to homogenize the metallurgical structure, remove stresses, or achieve the required surface finish.

The first basic functions were already described by the science fiction author Arthur C. Clarke in 1964. In 1987, the first stereolithography printer was presented. Starting in the 1990s, many technologies arose including fused deposition modeling (FDM) for plastic material and selective laser sintering (SLS) for prototyping. It took until 2009 when market introduction of desktop printers was achieved. Metal printers also appeared in the market and they have been showing high and increasing demand since then. In addition, related businesses appeared focusing on material development, powder production, post-heat treatments, surface finishing technologies, and many more.

The possibilities of metal 3D-printing increased from being a prototyping method to direct printing of parts, for repair or even to add functionalities to conventionally built structures. Due to decreasing costs for the AM machines, heat sources, and feedstock materials, the cost per additive part decreases and it becomes more attractive to produce additive parts compared to conventional subtractive or formative processes. The main drivers to use AM are:

- Possibilities to produce advance geometries including lightweight designs,
- Minimizing waste production, less inventory, shorter supply chains,
- Part consolidation, combining multiple parts into few, and
- Mass customization.

The AM procedure includes many steps from different disciplines, which makes the industrial implementation more complicated. It takes time until all experts agree on partly revolutionary new procedures and work together. Usually, it is not possible to produce an additive part the exact same way that a conventionally manufactured part appears. There is usually a new design needed. Depending on the printing method, support structures may also be needed to enable printing. Then, there is pre-processing needed, e.g., to digitally slice the part into different layers and create the G-code for machine movement to create the part. Often, printing requires process knowledge and adaption of process parameters. After printing, the material must be typically heat-treated either for binder removal, homogenization of the microstructure, or stress relieve. Sometimes, the heat treatment can or must be done during the process. In order to achieve the desired surface finish, local machining of surfaces is often needed. Layer heights and material properties vary depending on the used method and materials.

AM is categorized in seven main categories. The seven principal categories have multiple sub-categories and process variations.

1. *Material extrusion*: The material application for extrusion-based 3D-printing works similar to inkjet printers, while a thermoplastic filament from a coil or plastic pellet is fed into a heating system to liquefy. The plastic is then extruded through a nozzle. By moving the nozzle and table relative to each other, parts can be built layer by layer. Support structures are needed when building overhang elements with large angles. Such support structures can be built using a second extruder nozzle with a separate support material to apply material in the later hollow parts after its removal. This enables the printing of the final structure onto material in every layer, even when the final part has left out sections. Other possible materials that can be used are food or even live cells. Good material properties can be challenging to achieve. The applied material usually shows gaps and voids between layers and tracks. Therefore, tensile strength can be sufficient along the track direction but can be very low perpendicular to it. Often crisscross raster material application is used to minimize heterogeneity of properties in different directions. Due to thermal treatment, internal thermal stresses can occur during the process. Local thermal extension is followed by thermal shrinkage. If the shrinkage induces too high stresses in the material, the material can bend upward and lead to warping. The print must be interrupted. The process produces little waste and material changes are easy by

just changing the coil or granulate material. Due to the low accuracy and surface finish, this process has some disadvantages to be used for critical parts. In order to increase bonding strength and reduce anisotropy, layer heating is done just before applying the next layer that can improve the bonding. A laser beam can be used as a local heat source for pre-heating. The part strength can be improved using the local layer pre-heating.

2. *Sheet lamination*: This process is based on adding thin metal sheets layer by layer. The older method is based on gluing sheets to each other, which is called laminated object manufacturing. Ultrasonic consolidation is the more common process today. The clamping happens using a sonotrode, pressing the sheets together. In addition, the sonotrode rapidly moves with ultrasonic vibration, which induces the necessary heat to join the material sheets. Due to surface waviness and roughness of the sheets to be connected, the bonding can show voids and gaps that can reduce the mechanical properties. Due to relatively high costs and comparably high effort for post-treatment to achieve the final shape, this method became less popular.

3. *Binder jetting*: For binder jetting, pre-placed powder is necessary, which is prepared layer by layer. The structure is created by jetting a plastic binder over the powder layer to locally glue the powder particles together. The powder material is typically metal but ceramic or even sand can be used. The printing process is done at room temperature, which reduces the impact of thermal issues during printing. However, after drying the binder, the binder needs to be removed later in the post-processing. The related shrinkage during binder removal needs to be considered during the design phase to achieve the desired geometry and tolerances. Parts are cured in a furnace to become the green part before removing the powder material around. This is followed by de-binding of green parts and sintering, where the part shrinkage occurs, to reach the final part. Binder jetting is an upcoming AM technology. More research and understanding are currently necessary before a potential larger implementation in industry. Limitations in part size, material availabilities, and low mechanical properties are to be overcome.

4. *Vat photopolymerization*: A liquid container of resin (thermosets) is used as feedstock material. Local illumination with, e.g., an ultraviolet laser induces polymerization, a chemical reaction, of the photo-sensitive resin and the material becomes locally solid to form the final part layer by layer. Plastics and some ceramics are possible to use. Three main process variations exist: (a) The vector scan or point-wise technique uses a scanning laser to create the layers, (b) mask projection can create one whole layer with just one illumination step projecting the individual pattern to be solidified in one layer through a mask using a laser beam or lamp source in combination with an optical element, and (c) the two-photon approach uses the interference between two laser beams to locally induce the necessary heat for the chemical reaction. The process is relatively fast and accurate with good surface finish at a high degree of automation. Support structure design and

removal and the correct resin selection limit the possibilities to some extent. Material properties are moderate and can even degrade with time.

5. *Material jetting*: Using the principle of 2D inkjet printers, materials can also be printed in the third dimension by particle deposition. Deposition can be achieved using continuous inkjet deposition or particle on demand techniques.

6. *Powder bed fusion*: The structures are built in a powder bed of pre-placed powder particles. Plastics, metals, and sand can be used as feedstock. A laser beam or an electron beam scans over the powder layer to locally sinter or melt the powder. For plastic sintering, surface tension, viscosity, and particle size mainly define the sintering kinetics. Low viscosity and low particle size encourage the sintering process in general. Challenges occur due to thermal impact during the process that can lead to warping and material shrinkage. During plastic sintering, no support structures are needed in contrast to the metal melting process. Little post-processing is typically required if surface finish is not the main quality factor. Powder reusability is still a big challenge due to limited knowledge about the degradation of powder, impact of spatter, and accumulation of powder particles during the process.

7. *Directed energy deposition*: Metals and some ceramics are possible to print by blowing powder into a created melt pool. The energy source is typically a laser beam or an arc source.

3.12.1 PRINCIPLES OF LASER ADDITIVE MANUFACTURING

In this book, the focus of additive methods will be on the laser-based approaches for metal printing. The main principle of AM is the iterative application and melting of material, usually layer by layer. Metal-AM is commonly done using powder or wire as added materials, while a heat source creates a melt pool in which the material incorporates. Thereby, complex structures can be produced that contain, e.g., inner structures, overhangs, or undercuts [Bru17]. Every part can be practically customized. In the design stage, the specific additive requirements need to be considered in order to make the part printable. AM enables new designs possibilities, while at the same time, the design needs to be adapted in order to use the advantages of the new technologies and consider machine and processing constraints. Thereby, components that enable the combination of several parts can be manufactured in one component and lightweight designs are possible. In addition, multi-material designs are possible by applying desired material mixes. Typical materials used for AM are Titanium alloys, stainless steels, or Ni- and Co-based alloys.

After the computer aided design (CAD) model development, the part is sliced into layers for fabrication (computer aided manufacturing, CAM). In order to translate the design information into machine movement, digital twins are used. Advanced modeling software is able to include machine and part digital twins. Machine digital twins can simulate the movement during the process and enable the detection of incorrect robot or CNC-machine movements. In addition, the detection of collisions is possible to identify during the design and path planning phase, which can avoid expensive or even dangerous situations in the real world. When inserting the part to be produced in the digital model, the exact path can be visualized, and potential mistakes can be identified before printing by experienced users. Detailed knowledge about the machine and material behavior is necessary in the design phase, e.g., the shrinkage and wetting of the material needs to be correctly included in the path planning to avoid failures due to wrong nozzle-part-distances due to mistakes in layer height assumptions. Advanced slicing software is already available that provides adaptive path planning based on in-situ measurements of the actual height and can add or remove layers during the process and still achieve the designed part. After the design and slicing of the part, the machine-readable code is produced and sent to the machine, where the printing happens. Sensors can increase process stability and can detect failures. This is particularly of interest when building large parts. Either the production can be stopped when the failure occurs, saving time and money to print the entire part or the operator has the possibility to repair the failure before continuation of the automatic program.

The big advantage of AM is that geometric complexity can be added to the design at no extra cost. There is no tooling required (little starting costs) compared to, e.g., casting (molds are expensive). Reduced material need can be achieved by the new design possibilities, where material is placed and added only where needed from the load constraints of the structure. Repair applications can enable the reduction of metal waste, while often, the added structures show even higher performances than the original part. The main advantage of additive processes compared to subtractive methods is that there is a minimum waste material produced since the material is positioned and built where needed, while unused powder can (partially) be reused.

The main processing quality criterion is the production of defect-free volumes. However, porosity and material losses due to spattering can occur during processing and need to be avoided. Another challenge in multi-layer processing are the multiple thermal treatments of the volume elements in the structure that lead to complex microstructural developments and locally varying material properties.

For metal printing, both SLS and melting-based approaches can be used. For the sintering approach, a multi-component system is applied consisting of a high melting metal component in the form of powder and a low melting binder. During the process, the binder melts and the non-melted metal material sinters. Most applied metal additive methods based on laser illumination are the directed energy deposition (DED) and laser powder bed fusion processes, which are based on material melting and adding material to a melt pool.

3.12.2 DIRECTED ENERGY DEPOSITION

DED, also known as direct metal deposition (DMD), direct laser deposition (DLD), or laser metal deposition (LMD), uses powder or wire material for local incorporation into the laser-induced melt pool and thereby iteratively building structures. In this chapter, the DED process with laser beam for metals (DED-LB/M) will be described as one of many DED processes. This technology finds its main applications in building large structures or in repair works.

3.12.2.1 Processing Setup

Material is applied in the form of powder or wire, while the laser beam energy melts the additional material and partially the base material to create bonding. The material feedstock is typically applied continuously. The principle is close to a welding or re-melting process, while adding material into the created melt pools. Dense coatings, which are metallurgically bonded to the substrate can be achieved. When applying multiple tracks next to each other and on top of each other, complex 2D and 3D structures can be created. A large spectrum of materials can be used. Typical examples of powder materials used in DED are titanium, Inconel, stainless steel, copper, and aluminum alloys.

Different material application systems are possible. Lateral nozzles are easy-to-build and to use. Therefore, such a setup can be cheap, but is direction-dependent. This means that when building in different axial directions, the powder flow is applied onto the melt pool from different angles, which can influence the powder incorporation and thereby the track quality and height.

Therefore, usually coaxial nozzles are used for powder-DED (Figure 3.54a). They enable the direction-independent powder application and more efficient powder incorporation. The powder is guided and focused onto the material's surface by a powder nozzle. Complex inner designs of powder guiding inside the nozzle have been developed in order to provide a radially homogeneous powder flow. The laser beam is guided through the center of the nozzle and produces the melt pool in which the powder particles incorporate (Figure 3.54b). In many nozzles, a protection gas flow is used in addition to the transport gas flow that guides the powder. This additional shielding protects the process zone from oxidation.

The achievable building rates and related building speeds depend on several factors. The amount of material that the powder feeding system can provide through the equipment can be a limiting factor. In addition, the size of powder or wire material and the properties of the laser sources used can limit the achievable build rates. The laser power can be set to over 20 kW for high deposition rate processes with large spot sizes (over 10 mm in diameter). Other processes require only a few

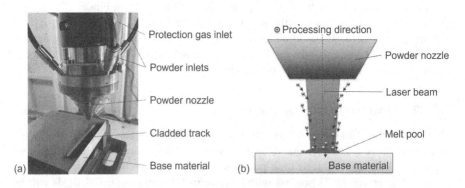

FIGURE 3.54 (a) Setup for blown powder processing and (b) sketch of the powder-DED process.

hundred Watts for precision deposition processes using small spot sizes (few hundred μm) [Dem18].

The DED process requires several equipment to enable the processing. A laser system is needed, including fiber optics, to control and guide the laser beam to the processing zone. The feeding nozzle transports the powder or wire. Coaxial nozzles also leave the space to transfer the laser beam through its center. The powder or wire feeder transfers the feedstock material to the nozzle. Powder feeders typically use filled hoppers that release the powder onto a rotating table with a groove. From this groove, the powder is collected, and the transport gas leads it through the tubes to the powder nozzle. Multiple hoppers in combination with multiple inlets into the nozzle enable the mixing of different powders for in-situ alloying or creation of graded structures. A CNC or robot system enables the relative movement between the optic with nozzle and the part. Typical particle speeds used to transfer the powder particles are between 1 m/s and 4 m/s [Pra18].

3.12.2.2   Laser Energy Absorption and Material Interaction

Laser beam absorption during the DED process is still difficult to predict. Depending on the laser beam size and the induced size of the melt pool, the laser beam can illuminate the powder material, the liquid melt pool, the structure underneath, or the solidifying track. All these surfaces have a different surface roughness and atomic structure. Therefore, the absorption is different. In addition, absorption varies with temperature as well, while different dependencies can occur. The temperature dependency of most materials is still unknown, in particular, at high temperatures as they are reached in metal melt pools. Since the melt pools and also the resulting track surfaces show rounded shapes, the angle-dependence of the absorption needs to be considered as well. Although typically unwanted, local vaporization can occur on the melt surface, which can induce dynamic melt pool behavior leading to spattering or pore formation or even reduce the laser light transmission due to laser light absorption in the vapor.

In general, the knowledge about material properties of the used materials for DED is essential but is still not fully available. The material defines not only the absorption of laser energy, but also thermal conductivity of the heat being transferred. The higher the thermal conductivity, the less energy is available to create the melt pool, e.g., when depositing copper powder on copper base material, the melt pool is comparably small due to the excellent heat conductivity of the copper base material. Copper applied on steel substrate shows larger melt pools [Pra20]. The material also defines the melt pool size, melt flow, and the incorporation behavior.

3.12.2.2.1   DED Materials and Transfer

The diameter size range of powder materials used for DED processes is typically between 50 μm and 200 μm. The processing results can vary depending on the used powder size distribution and powder feed rate (Figure 3.55).

The powder is transferred through powder nozzle, while the powder stream focusing depends on the powder nozzle and powder material. For efficient powder delivery and the avoidance of clogging the powder feed system, good powder flowability is required, which makes the use of spherical powders preferable. However, the use of

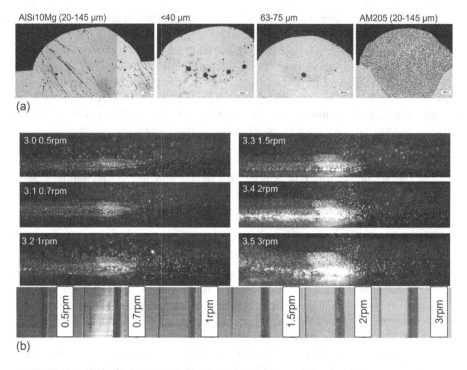

(a)

(b)

FIGURE 3.55 Cross-sectional views of single-line DED tracks at different (a) powder size distributions and (b) powder feed rates.

used powder containing, e.g., agglomerates is explored to a great extent in order to minimize waste production and enable efficient material re-use. Often, powder can be re-used for applications with lower requirements on the chemical composition or part performance. However, used or recycled powder can influence the process behavior and process outcome, in particular, the formation of porosity (Figure 3.56).

The powder particles are released from the nozzle being directed toward the processing zone. In conventional powder DED, the powder particles only interact with the laser beam for a short period of time and are slightly heated but do not reach melting temperatures. Therefore, powder particles are solid when they arrive on the melt pool. There are process variants for high-speed cladding, where the powder particles are liquefied in a laser beam of high intensity and then sprayed onto the structure.

FIGURE 3.56 Cross-sectional views of DED walls with differently treated powder.

The deposition rates for the laser DED processes can vary depending on the process setup and deposited material from less than 100 g/h to several kilograms per hour [Pra20, Zha10]. High-speed laser cladding could achieve 2 kg/h [Zho15]. When using combinations of heat sources, like laser-arc-processes, build-up rates of 18 kg/h have been demonstrated [Now15]. The lateral resolutions cover the ranges between 30 µm and 5 mm, while the part sizes are, in general, not limited.

3.12.2.3 Impact of Process Parameters

Process parameters need to be adapted to the material used and the required dimensions. The energy input can be varied using laser power, process speed, or focus position. The laser power is one of the main parameters to alter the laser beam intensity (Figure 3.57).

The process speed defines the energy input per time or distance (Figure 3.58).

Together, the laser power and process speed define the line energy input. A high line energy leads to larger and deeper melt pools. In general, a faster incorporation of powder particles can be expected due to more rapid powder heating when they landed on the melt pool and a reduced surface tension of the melt pool at higher surface temperatures. Low energy input can lead to lack of fusion and production of voids. The powder feed rate mainly defines the height of the tracks produced. However, indirectly, the depth of the clad track is affected since the applied laser energy is needed to melt the powder, which reduces the available energy for dilution. A laminar flow from the powder nozzle is typically desired to avoid oxygen contamination of the processing zone. Both too high and too low flow rates can induce turbulent flows leading to an increased oxygen content around the melt pool. Another important parameter in DED is the hatch distance. The hatch distance defines the overlap between the applied tracks. A certain overlap is needed to create a proper structure. Too low distance between the tracks, or small hatch distances, can lead to wavy occurrence of the applied structure, while too high chosen hatch distances can lead to melt accumulation due to the strong surface tension effect leading to balling of the added tracks. Often a good starting point is a 50% hatch distance to create surfaces with minimum waviness.

3.12.2.4 Powder Particle Incorporation

Figure 3.59 shows high-speed images of the melt pool and the incorporation of powder particles. The transported particles from the nozzle are directed toward the melt pool. It should be aimed to have the powder focus positioned on the material melt pool surface in order to make particles hit the molten material. Some particles will collide during the path from the nozzle to the melt pool and change direction. They might be directed to solid surfaces such as the base material structure or the solidified track, and will be lost for the process. In addition, the loose particles need to be caught by an exhaust system to avoid air contamination and humans inhaling such particles. The catchment efficiency of the process can be high up to 98%. This is possible, when the melt pool is large enough, particles are correctly directed onto the melt pool, and the melt pool shows minimum oxide coverage that might avoid the incorporation. Most steels show a limited oxide layer appearance on the melt pool surface, while, e.g., on aluminum alloys, continuous coverage is possible that limits

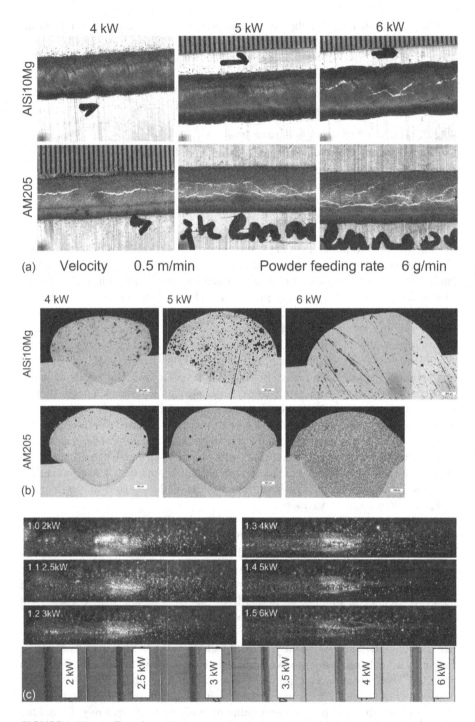

FIGURE 3.57 (a) Top view, (b) cross-sectional view, and (c) high-speed images of single-line DED tracks at different values of laser power and aluminum alloy.

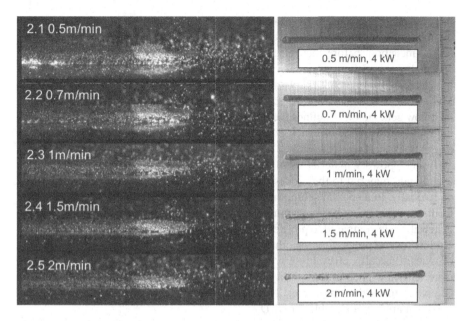

FIGURE 3.58 High-speed images and top view of single-line DED tracks of aluminum.

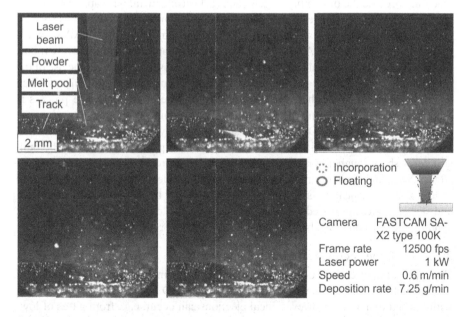

FIGURE 3.59 Sequence of high-speed images of a blown powder laser deposition process.

the amount of material to be added. It can be seen that all particles that land on the melt pool are incorporated. Depending on their landing zone, the particles float on the melt pool before incorporation [Vol18b]. The densities of powder material and melt pools define the incorporation duration and characteristics. Some alloying elements can increase the wettability, and a rapid incorporation is possible. The Marangoni flow induced by the temperature-dependent surface tension is the main melt flow generator and defines the floating path of powder particles. In case of landing in the mushy zone of a melt pool, the particles can remain as surface roughness on the track.

Besides powder material as a feedstock material for laser-DED, a wire can be used to add material to the process zone. Wire can be fed by a lateral wire feeding or coaxially. In a coaxial setup, the laser beam needs to be guided around the wire feeding system using beam shaping optics. The use of a wire enables a reduction in material waste compared to powder feeding where powder particles can ricochet and are lost for the processing. Another reason of using a wire is the good availability of wires. Basically, all welding wires can be used for DED. However, not all wire materials show good processability in DED due to the more complex heat treatment compared to welding processes. Nowadays, wire materials are increasingly developed that are better suitable for DED.

3.12.2.5　DED Heat Sources

Laser beams are not the only possible heat source to be used for DED. Arc welding machines provide the possibility to generate tracks using metal wire application. The wire is melted by an electric arc and transferred into a melt pool. The process is very similar to arc welding but is called wire-arc additive manufacturing (WAAM) when used to build structures. WAAM is therefore a well-established and relatively robust process. Typically, the metal inert gas (MIG) process is used, along with the use of the tungsten inert gas (TIG) or plasma processes, e.g., when processing titanium. When using the MIG process, where wire is used to produce electric arc, the process is direction-independent. Using a robot system, very complex structures can be produced. Compared to laser-DED, WAAM shows typically higher deposition rates and lower investment due to relatively cheap heat source equipment. However, the built structures show rougher surfaces, which require post-processing like milling of the edges. Arc sources also introduce a comparably high heat input, which can lead to distortion and overheating issues of the structure. Wider tracks are usually produced, and slower processing speeds need to be used. Similar to welding, DED can also be conducted using combined heat sources of laser and arc, resulting in a hybrid process.

3.12.2.6　Imperfections and Monitoring

The typical imperfections occurring during DED are similar to the heat conduction welding defects. Since no keyhole is produced, the DED melt pools are usually calm with limited dynamics. However, local ejections can occur, e.g., from gases of low-boiling elements that form in the melt pool and escape, local overheating and ejection of oxide skin parts, or sudden melt flow accelerations. Such spatters can induce a material loss and even damage equipment. In addition, spatter can land and solidify on the previously built structure and form unwanted surface structures.

FIGURE 3.60   Sheet deformations at single-line DED track deposition at different values of laser power.

As in most thermal processes, hot cracks can occur during cooling in the mushy zone. They can form along or perpendicular to the track deposition direction and depend on the material properties. Residual stresses induce such cracks supported by alloying elements that form intermetallic compounds. Oxide skins can reduce the bonding strength between layers [Hau21]. The heat input can lead to deformations (Figure 3.60).

Due to the complexity of the process, results can vary, which makes the process monitoring relevant and partly mandatory in order to achieve the part quality requirements. Until now, process monitoring is often used for the detection of irregularities and a manual adaption of the processing. Control systems are increasingly developed that can alter the processing parameters when certain sensor data detect irregularities. Immediate adaptions of process parameters during the processing can directly heal a pore or lack of fusion. Process parameter adaption for locally adapted energy input during the next layer can compensate for the defects as well.

The processing zone offers multiple emissions that can be used for monitoring and control. Acoustic emissions provide information about the melt pool dynamics, while light emissions contain melt pool and track dimensions as well as temperature data. Even vapor emissions can give indications about the process stability.

- Acoustic emission measurements have started to become more popular for process monitoring. The challenge of typically noisy environments needs to be solved or considered in the measurements to avoid false alarms. Airborne or structure-borne acoustic emissions can be recorded using microphones [Hau22].
- Light emission detectors are often used. Visual cameras can detect and monitor the melt pool dimensions over time and warn about dimension changes that can indicate defects. Increased dimensions can, e.g., indicate overheating due to a lack of fusion. Due to high costs, high-speed cameras are common for research purposes and give insight into the melt pool

behavior, spattering, and vapor characteristics. Spectral analysis offers the possibility to detect temperature and radiation intensity peaks that can be related to certain ionization lines of elements. Thereby, e.g., the composition of the vapor can be monitored [Vol23a].

- Optical coherence tomography has become more attractive for monitoring purposes [Kog20]. High-frequency recording of distances enables the online detection of height variations. This can be used for melt pool height monitoring. When scanning the measuring system, the pre-measurement of the appearance of the structure before applying the next layer can be detected, which enables the adaption of the path planning in-situ. For quality control purposes, topology of the solidified track can be performed.

- Thermal monitoring becomes very relevant in additive processes due to intrinsic and partly complex thermal developments. For a sufficient heat management, thermal process data are needed. Thermal data can indicate melt pool behavior and even pore formation or lack of fusion to the layer underneath. When heat dissipation changes, the temperature field on the material's surface will change.

Sensors can be mounted in different ways. Height detection systems or thermal imaging cameras are typically fixed off-axis, while visual and some thermal imaging can be done coaxially.

Production is still organized in the way that quality control of parts is done after the production process. Either selected or even all produced parts go through quality control and testing. In AM processes, it is now also possible to monitor and check the quality during the process. Such procedures take some time until implemented but show high potential for saving quality control time. Current research efforts aim to make the monitoring and control systems reliable and robust. That way, control systems can be developed to ensure part quality by process monitoring and in-situ corrections. For detection of defects in sensor data and enabling control loops with rapid decision-making, machine learning algorithms can be useful to include.

After completion of the part production, often post-processing is necessary to achieve the final part shape. For post-processing, advanced machines offer the possibility to process the additive part in the same cabin to produce the final net shape using subtractive methods [Kar10].

3.12.2.7   DED Applications

Typical applications of laser DED can be found where functional features or structures are needed to be added to the existing parts in order to avoid machining of larger blocks. Furthermore, in the field of part repair and remanufacturing, DED shows advantages and high potentials, in particular, with regard to limited available material sources and sustainability goals. In particular, high-value material repair shows the highest user potential and economic benefits. DED also offers the possibility to place different materials in the same part by changing the feedstock material. This can be done, e.g., using multiple wire feeders or several hoppers for powder feeding or combinations of both. Due to typically rough surfaces of the built structures, often post-processing is needed in the form of machining selected surfaces to achieve the required surface finish. In addition, heat treatment of the material might be needed to relief stresses and homogenize the material due to possibly occurring undesired local microstructural developments during multiple heat treatments.

3.12.3 Laser Powder Bed Fusion

Powder bed fusion (PBF), also known as selective laser melting (SLM) or laser metal fusion, is a powder bed-based additive technology. The PBF process with a laser beam of metals (PBF-LB/M) is described here.

3.12.3.1 Process Principle

The pre-placed powder is locally melted and fused to the structure underneath. By applying powder layer by layer and local melting of the material, complex structures can be produced (Figure 3.61).

Here, 3D CAD-models of the parts are needed. The design must fulfill several criteria to be used for building in PBF machines, e.g., support structures are needed when building complex overhang elements. After design, a digital part needs to be virtually sliced in layers. This information is used by the machine to calculate the laser paths for each individual layer.

After application of the powder layer of a defined height, the laser locally melts the material, before the powder bed is lowered, and the powder application starts again. This procedure is repeated until the whole part is produced. The powder sizes used for metal PBF are typically in the range of 20–60 μm. These particles are

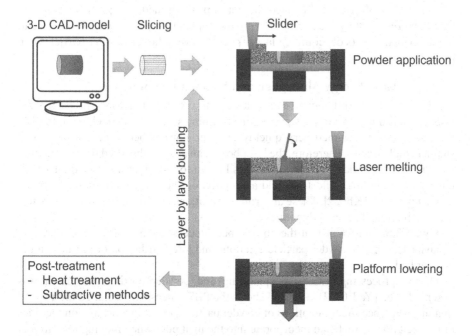

FIGURE 3.61 Principle of automated powder bed additive manufacturing (AM).

very small and can cause damage in human bodies when entered the lungs. Some of those particles can even enter the body through skin pores. Therefore, sufficient safety measures are needed to handle the powder in a safe way. In particular, alloys containing hazardous elements should be treated with care. For better spreadability during the powder layer filling, spherical powders are usually preferred. However, these powders are often expensive, and the use of non-spherical powders is under investigation.

Powder layers are automatically applied by most industrial machines. Thereby, the build plate is lowered, and new powder is added on top of the previous layer. The build plate lowering can be defined by defining the layer height. However, it needs to be considered that the actual height of the produced tracks is not the same as the programmed layer height. Due to powder material loss as well as track wetting and shrinkage, the track height is lower than the previous powder layer height. Due to melting of powder particles, gaps between the powder particles disappear, which reduces the volume. Therefore, the powder layer height varies between the layers. The very first layer has exactly the programmed height. After processing the first layer, the build plate is lowered but the new layer height plus the distance from the track surface to the intended first layer height needs to be filled as well. Theoretically, every layer should be programmed with different laser parameters to account for the different amount of available powder. This is typically not done in actual PBF machines, and conservative parameters are used to guarantee the powder melting even when more powder is available. This variation also explains some machine producer recommendations. Sometimes, the powder layer height is recommended to be smaller than the larger particles in the powder size distribution, e.g., if 40 μm powder layer height is programmed, the particles >40 μm should be removed by the scraper during the application of a powder layer. Due to the described effects of requiring larger powder layers, larger powder particles can be used as well.

3.12.3.2 Laser Energy Absorption and Material Interaction

The laser light absorption can be complex due to variation in surface properties the laser beam illuminates. In DED, the main absorbing surface is the melt pool. In PBF, the laser beam is scanned very quickly over the powder bed, which forms much smaller single tracks. Therefore, the laser beam can also be absorbed by the structure in front or the powder. Powder material is known to absorb more laser light due to the effect of multiple reflections and absorption on the surface consisting of many powder particles [Kho22]. These absorption variations can be one cause for the varying heat input and process dynamics.

Figure 3.62 illustrates that the powder bed processes show an inherent dynamic behavior leading to powder particle ejections and a redistribution of powder on the powder bed.

During processing, extensive spattering and powder particle movement was observed (e.g., [Yad10]). Figure 3.63 shows the PBF process in a detailed high-speed imaging sequence. Accumulations of powder particles occur in front and on the sides of the melt pool, which can incorporate into the melt pool when moving forward but can also remain on the material's surface. Thermally induced gas flows can initiate

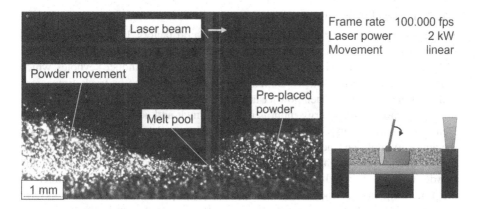

FIGURE 3.62 High-speed image of a powder bed fusion (PBF) process.[1]

the powder particle movement on the powder bed surface toward the melt pool. This denudation effect leads to a redistribution of powder and varying powder amounts for the subsequent tracks in the actual layer.

Material ejections from the processing zone may involve non-melted particles that are accelerated by gas or vapor flow [Bid18] or melt ejections. This effect can also lead to a redistribution of material on the powder bed surface. Since accumulated powder particles are partially ejected, not only the amount of powder particles on the powder bed surface can vary but also the size distribution can differ. In addition, the comparably large agglomerates can lead to difficulties for the scraper to add the next powder layer. Both agglomerates and spatters that land on the powder bed can show different properties. Laser light absorption of, e.g., oxidized particles can vary, leading to variations in energy input and increased process dynamics. In addition, the previous heat treatment can lead to unwanted oxide inclusion into the subsequent tracks when those particles are incorporated into the next tracks and layers.

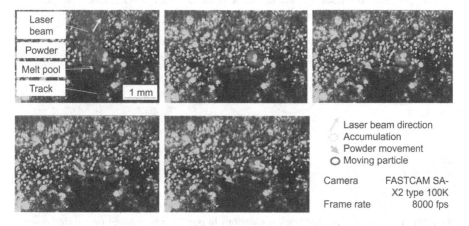

FIGURE 3.63 High-speed image sequence during powder bed fusion.

3.12.3.3 Track Appearances

The geometry of the single tracks depends on several factors. In general, the tracks show cross-sectional shapes comparable to heat conduction welding tracks. Some PBF machines operate at higher laser beam intensities and can induce keyholes, which lead to deeper and wider tracks. Depending on the conditions, the track dimensions and positions can vary. When powder is available, the tracks appear at a higher position, since most of the energy input is needed to melt and add the powder material. Denudation effects can lead to variations in powder availability. Denudation during the application of one track reduces the available powder around the track. This can happen due to induced gas and vapor flows during the processing, which induce the powder to move. It could be shown that the powder material availability influences the processing and the track dimensions [Vol19b]. It is possible that some tracks have no powder available to incorporate. These tracks appear mainly in the base structure and just re-melt the already processed part without adding material. Therefore, the track dimensions can vary within the same layer (Figure 3.64). Of course, such energy input without adding material is inefficient and should be avoided. However, re-melting can also heal defects that occurred in previous layers, e.g. porosity.

These variations usually lead to the use of conservative processing parameters in order to avoid inclusions, lack of fusion, and porosity. The high-energy input has been shown to lead to several re-melting cycles of the same material when building a structure [Mis18], which increase the energy used for the process. It has been seen that the same material can be re-melted up to seven times with typical process parameters.

In order to avoid irregularities, different laser beam scanning methods are used in powder bed processes. Thereby, the scanning direction can be changed in each layer, e.g., linear in one direction, zig-zag, or in spirals [Win17]. Xu et al. [Xu17] demonstrated, e.g., uniform 5 × 5 (tracks × layers) block samples. The change in such processing directions in each layer can reduce the influence of directional defects or inhomogeneous material properties.

Another important factor is the overlapping of the single tracks. This hatch distance has been found to induce variations in the track geometry and surface morphology

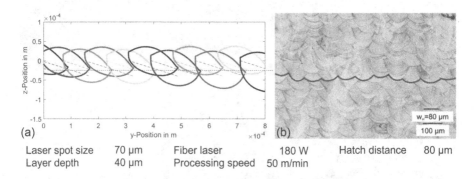

| Laser spot size | 70 µm | Fiber laser | 180 W | Hatch distance | 80 µm |
| Layer depth | 40 µm | Processing speed | 50 m/min | | |

FIGURE 3.64 (a) Calculated track dimensions in one layer depending on varying powder availability and (b) cross section of an additively built cube [Vol19b].

[Yad11]. Also, the scanning speed and power density of the laser beam impact the final metallurgy, part geometry, and accuracy. Sufficient energy is required to melt powder particles and enable bonding to the previously built structure, while excessive energy input can lead to turbulences and heat accumulations. For comparison of energy input, in PBF-LB, the energy density is often used relating the laser power (in Watt) to the scanning speed (in millimeter per second), hatch distance (in millimeter), and layer thickness (in millimeter).

3.12.3.4 Material Properties

The material and part properties after printing depend on most material and process parameters. Part properties can significantly vary when using different build directions and energy densities. Material properties are defined by cooling conditions and multiple thermal cycles. Laser-related parameters are primarily the laser power (typical range from a few Watts to 500 W), laser spot size on the material (around 80 µm), and pulse parameters (pulse duration and frequency if pulsed lasers are used). Scanning parameters are the scan speed, hatch distance, and the used scan pattern. Powder type and material properties are main influencing factors on part appearance. Particle shapes, sizes and size distributions, the powder bed density, and the layer thickness need to be optimized for successful processing. Furthermore, the pre-heating temperature of the powder bed can influence the processing and resulting material properties.

3.12.3.5 Powder Recycling

In order to provide high-quality parts, often virgin powder is preferred for production. However, the used powder can be re-used. Due to the thermal character of the process, the powder material that does not become part of the structure can be thermally influenced. Also spatters or agglomerated particles can remain in the powder. Since oxidation can happen during heating of powder particles and microstructural changes can be induced, the characteristic of the used powder is often not completely clear. Standard procedures often include the mixing of used and fresh powder for a few times until the powder is considered to be waste. Both the effect of different heat treatments of powder particles and the impact of those particles on the process and part quality are ongoing research fields. It has been shown that used powder contains more oxides, which are transferred into the melt pool and can increase the melt pool oxide coverage, leading to reduced material incorporation and increased defect occurrence.

3.12.3.6 Process Monitoring and Defects

Process monitoring can help to limit defect production. Most industrial machines have multiple monitoring systems integrated, e.g., the gas composition and flow in the build chamber are well-controlled, which avoid many contamination effects and transport spatter immediately away. Often, the build plate temperature is controlled as well to maintain constant conditions during the building.

However, in order to get immediate feedback on the track quality, additional sensors are needed. Some systems allow the coaxial visual observation during the processing. A visual camera can observe the melt pool during the processing and detect variations in its dimensions. Thermal imaging is highly desired to detect overheating or defects that induce surface temperature changes. Such systems are typically more difficult to integrate into industrial systems at the moment. Thermal measurements can also be conducted between the processing of the layers to check for unexpected temperature fields of the whole powder surface. These systems can also be integrated as coaxial systems.

In order to alter the process and minimize defect creations, beam shaping methods are increasingly investigated. Certain beam shapes offer a more homogeneous energy input and potentially less spattering and denudation effects. Some PBF machines allow the integration of beam shaping equipment into the laser beam path. However, the field of beam shaping for PBF is under investigation.

3.12.3.7 Systems and Post-Treatments

The industrially available machines use similar parameters for processing. These are based on applying comparably small powder size distributions and illuminating with low laser power ranges using fast scanning. However, it has been shown that PBF-LB/M at very high laser power is also possible, e.g., at 16 kW [Lei21]. Such new ideas and process variations can support the flexibility and range of applications for AM methods in the future.

For some applications, post-treatment using subtractive methods is necessary to achieve the final part shape. Due to the extensive and repeated heat treatment, post-heat-treatment can be necessary.

Although PBF is widely used and has been proven to be a promising technology to build parts and structures of complex designs at high build-up rates or high precision, the process must be further understood and optimized for increased material and energy efficiency. Compared to DED, in general, PBF shows higher accuracy at lower building rates.

3.12.4 POWDER SHEET ADDITIVE MANUFACTURING

In the field of AM, many process variations are continuously developed solving drawbacks of the original idea with advanced technical ideas, e.g., the process duration can be decreased when using multiple laser heat sources in PBF-LB/M at the same time.

One variation in laser-based AM is the MAPS (metal additive manufacturing using powder sheets) approach [Lup22]. The process uses powder sheets instead of loose powder as feedstock. Since powder particles used for metal powder AM techniques are comparably small, precautions are necessary to protect human health. When using powder sheets, the powder is fixed in the sheet and the workers are not exposed to the powder.

In order to create powder sheets, powder needs to be embedded into sheets. Therefore, a polymer binder is used. The powder particles are pre-mixed with the binder and a solvent. This way, the slurry can be deposited on a flexible teflon sheet to form the sheet. The solvent quickly vaporizes and the binder acts as a glue to join the powder particles. This is an additional process step in the powder feedstock production but is expected to add minor costs. The advantage of the provision of the powder material incorporated in the sheet is that the powder flowability plays a minor role in the delivery of powder for laser processing. Therefore, nearly all kinds of powder size distributions and powder shapes can be used. During powder atomization, lots of efforts are needed to create the desired powder size distributions. Often, the powder sizes that are produced in the atomizer and do not fit the size requirements are sieved and even denote waste. The integration into powder sheets offers the possibility for increasing the ranges of powder size that can be used for AM processes and reduce waste. In addition, there is the possibility to use non-spherical powders from cheaper atomization processes, e.g., water atomization instead of gas atomization [Fed20]. The sheets can be produced as long stripes.

The powder sheets show a specific appearance due to the production process. During the vaporization of the solvent, the binder sinks within the powder sheet and accumulates on the lower part of the sheet. Therefore, the bottom sheet side appears relatively smooth, while the top side is dominated by the roughness due to sticking out of the powder material. Using the powder or binder side up during the processing can influence the processing.

When powder sheets are produced, they are placed on top of the base material sheet, similar to adding a layer for the PBF process. When one layer is processed by the laser beam, the sheet is moved by a roller system to the next position to provide the material for the next layer. The principle is similar to sheet lamination processes with respect to feedstock provision. Two variants using the laser beam are possible. On the one hand, laser beam scanning can be used, similar to the layer production during PBF systems, which requires the programming of the scan pattern for each layer. On the other hand, a new variant opens new processing possibilities. When using larger laser spots at high power, thick sheets can also be processed. This enables the application of more material at a time and therefore higher build-up rates. In both cases, shielding of the process zone is recommended to avoid oxidation effects. When a laser beam illuminates the powder sheet, the binder vaporizes at low temperatures, releasing the powder particles. These are then molten and incorporated into the created melt pool.

As mentioned, the decision of using a binder or powder side being illuminated by a laser beam can alter the process. A larger vapor plume is created toward the illuminated powder sheet direction when the binder side is on top. Since the binder creates vapor, vapor pressure is created that needs to escape from the processing zone and can eject powder particles as well, which are then lost for processing the additive track. Binder residues have been found on the powder sheets after processing [Vol23f]. These are condensed binder material from the laser illumination. In general, the powder sheets can be easily re-used. Adding solvents enables the formation of slurry again and a new deposition of sheets. The impact of the remaining binder residues might need to be considered when re-using the powder sheets with respect to their possible impact on the processing.

3.12.5 PART DISASSEMBLING USING LASER BEAMS

Parts and components produced with additive methods become more complex. However, the complexity is not limited to the geometrical shape, moving elements, or inner structures. Multi-material designs become more feasible, and many material combinations are possible. Due to the nature of AM having access to basically every

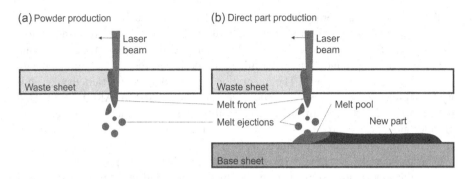

FIGURE 3.65 Sketches of the direct recycling approach used (a) for powder production for easy separation of multi-materials and (b) for direct part production from waste sheets to avoid recycling steps.

volume element, different materials can be integrated into parts. This can help in reducing weight by just having more material or material of higher strength at critical sections or enabling a graded transfer of properties using graded structures from one material to another.

However, when using multi-materials or even dissimilar materials in one part, the disassembling after the part's lifetime becomes challenging. When dissimilar materials are used, the materials end up in the same recycling route, which can lead to contaminations that are difficult to separate. In order to enable the separation of additive parts before sorting and melting, a laser-based process has been developed that basically enables "reverse" AM [Sam17]. The CYCLAM (reCYCling by laser ablation to feed additive manufacturing) process uses a high-intense laser beam to locally create a melt pool and recoil pressure from vaporization similar to the remote cutting process. That way, material can be gained from a larger part or component by controlled ejections. Powder particles can be either collected for further use in powder-based processes or even be transferred to a new track for directly building a new additive part (Figure 3.65).

3.12.6 WHEN TO USE AM?

AM methods are beneficial when high geometrical complexity is needed and customized parts are in focus of manufacturing since every part can be produced in different ways. When the number of parts to be produced is high and the complexity of the parts is low, conventional methods are often the better economic choice. Some additive methods are limited to the maximum build volume. However, these building possibilities increase all the time, e.g., DED processes show the capability of producing large freeform parts.

The main drivers that push the AM technologies forward are as follows:

- *Design opportunities*: Engineers and designers value the possibilities of part integration/consolidation. Fewer parts are needed for building components, while less inventory is necessary. Faster assembly at shorter supply chains is possible, leading to reduced costs and lower chances of making mistakes. Lightweight design and the possibility to add features and functionalities enable improved products. Rapid adaption to changed demands is possible due to flexible processes.

- *Business opportunities*: From a business perspective, the reduced supply chain, reduced inventory, and reduced downtimes due to increased reliability can lead to significant cost reductions. In addition, typically less labor cost is needed for printing parts and less material needs to be purchased. The reduction in complex production methods limits the risk of failures and customer complaints also that can lead to costly replacement or even court cases. Flexibility also offers an access to very different markets using the same AM technologies.
- *Customers*: Customers benefit from individualized and customized parts and reduced costs of products.

In the medical field, AM shows large potential. For visualization, body parts, e.g., organs or arteria can be modeled, surgical tools can be optimized, prothesis can be customized to the person, and implants can be individually created. There are even biodegradable implants that disappear, e.g., during the regrowth of a bone, reducing the number of surgeries necessary.

The aerospace industry was one of the first drivers of AM due to comparably small number of parts to be produced and high level of customization of parts. Additively created satellite parts are already flying; engine parts are consolidated and made more efficient and lightweight.

The automotive industry started a bit later with AM, since for mass production often conventional methods are more efficient and economical. However, additive parts are increasingly produced with AM. The main challenge is the certification of those parts.

Prototyping and product development can benefit from rapid manufacturing and testing of parts with low costs. In addition, AM can be beneficial when special materials are needed that are hard to process other ways, e.g., high heat resistance materials or materials embedded with metal, ceramic, wood, or carbon fibers. A wide range of properties are possible to create, e.g., high strength, stiffness, or biocompatibility.

Additive processes can still take more time compared to conventional manufacturing methods. Therefore, AM is mainly used when high precision is necessary, designs are not possible to be produced with conventional methods, or high-value material needs to be processed. Processing speed can be increased, e.g., by using several lasers for parallel working on one powder bed.

It needs to be considered that constraints can occur regarding the direction dependence of properties due to the building strategy. The process is comparably slow and can show restrictions in mass production and high costs for the parts, although high quality can be achieved. Depending on the process and feeding material, the accuracy and tolerances can vary. Often post-processing is required, which is an additional cost factor.

Research efforts are in progress to learn more about the properties of the final parts that have been produced by additive methods. For industrial implications,

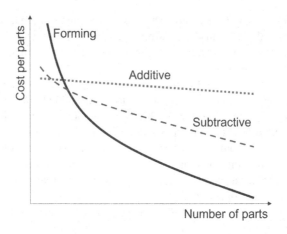

FIGURE 3.66   Comparison of costs per part with different manufacturing strategies.

relevant material analysis is often missing for a reliable prediction of the lifetime behavior of the part.

The cost of AM parts depends on the consumable time to print and post-processing (support removal, surface finishing). The cost per part for different manufacturing approaches is shown in Figure 3.66.

3.13   RESEARCH TOPICS

Research is conducted in most fields described in this book. Based on new advances in laser developments and process understanding, process improvements and new processing possibilities can arise. This last section gives a brief overview of the current research topics and developments without claim of completeness.

3.13.1   LASER DEVELOPMENTS

Pulsed laser developments are opening new possibilities. Ultrashort pulse lasers are getting more compact and easier to use, and integrate into production facilities. Industrial applications can be shock peening [Zan23], foil cutting, drilling [Sun24, Zha23a], and biomedical applications include micromachining medical stents [Cha21]. Advances in solid-state lasers such as the use of passive mode locking (e.g., [Ken23]) and robust pulse amplification systems enable highly reliable and low-maintenance systems.

Developments toward higher peak powers during the ultrashort pulses are ongoing to increase the ablation rate per pulse. Femtosecond systems compared to pico- or nanosecond pulse systems promise reduced thermal impact on materials, and reduced HAZ and defects. The limits and phenomena in these regimes need to be further investigated to enable processing of temperature-sensitive materials. Above the material-specific ablation threshold, saturation can occur, and plasma may absorb the laser energy of even the subsequent pulse [Zha23c].

However, new possibilities can arise, e.g., producing two- or multi-photon absorption in materials and manufacture of 3D parts [Gon23]. In marking applications, ultrashort pulse processing of metals can induce precise periodic structures in nanometer scale [Rat22] and enable more efficient and minimum impact on the surface, e.g., for medical tools. During welding, new possibilities to join advanced technical materials [Pri23] and brittle materials such as glass can be a potential new field for femtosecond laser processing [Cho20b].

3.13.1.1 High-Power Cw Lasers

The efficiency and maximum power of cw lasers are permanently improved and increased. Many effects of the increasing laser intensities on the material are still unknown and the potentials for materials processing need to be explored. For materials processing, the highest publicly reported material processing laser output power was 100 kW [Kaw18], while recently a 120 kW system has been installed [Rei23]. At high laser powers, simulations show that for aluminum, energy absorptance is constant at ~10% independent from the laser power, while for steel, the absorptance is simulated to decrease at higher laser power. Furthermore, new optical systems are in development that promise an increased laser power output of fiber lasers without losing beam quality [Che23]. Future investigations will show the underlying effects of such extreme laser-matter-interaction phenomena.

3.13.1.2 Lasers at Different Wavelengths

Advances in using different wavelengths compared to the typical ~1 μm wavelength are in progress to be established for high-power laser materials processing. Green lasers and frequency-doubled IR lasers are already established in the market and show advantages when processing, e.g., copper materials. Blue systems are increasingly developed. An all-solid-state 100 W scale blue-laser system has been shown to be stable and robust to be usable for precise copper applications [Jay24]. Such developments are based on laser-diode arrays. Current systems emit 40–100 W cw at $\lambda=452.2\pm2.5$ nm with 98% power stability and ~24% wall-plug efficiency, achieving multiple kilowatt per square centimeter intensities. Further advances in laser developments can open more possibilities for a more efficient processing and increased materials processing possibilities.

3.13.2 LASER PROCESSES

3.13.2.1 Process Developments

In DED-LB/M, the creation of the desired properties of a material is an ongoing research field. The control of alloying elements in combination with reactions with gas and temperature control is a topic of research for tailoring materials for AM, e.g., Zhang et al. [Zha23b] reported that Si-Cr-Mn oxides are formed in 316L process and have poor bonding with the deposited material. With a large size (0.36 ± 0.31 μm) and volume fraction (1.4%) of the oxides, the bonding strength of 316L is weak, leading to a low ductility. Adding TiC particles has been found to effectively change the oxides into uniformly distributed TiO_2 and decrease the size (0.24 ± 0.13 μm) and volume fraction (0.6%) of the oxides. With the refinement and decrease of the oxides, the bonding strength of the deposited materials is obviously strengthened, leading to a significant increase in ductility (25.4%, increased by 54.1%). Meanwhile, the yield strength and ultimate tensile strength have increased by 5.6% and 11.4%, respectively. DED offers many possibilities but the equipment is mainly used in production environments. However, a portable system for performing DED has been presented by Pang and Nellian [Nel22]. First successes have been reported of a repair application on rail steel components using Stellite 6 powder.

For PBF-LB/M, path planning optimization in combination with multi-laser machine coordination is a relevant current research field. An optimization strategy for laser scan paths is highly relevant to avoid, e.g., overheating or undesired temporal thermal developments. In PBF-LB/M, scan strategies are used in the layer-by-layer build-up of the workpieces that are pre-defined by machine producers. However, these are naturally not optimized procedures and can lead to heat accumulation and thermal stress as well as long laser off-times for multi-scanner systems. However, the optimization of scan paths for each layer is computational expensive and is difficult to solve on conventional computers within reasonable time. The recent rise of quantum computers promises to solve optimization problems in highly reduced computational time with increased solution quality. To exploit these advantages for PBF-LB/M, an adaptation of the multi-vehicle routing problem has been already presented [Bus23]. This procedure allows reducing the overall scan time while accounting for the inter-action of multi-scanner systems, reducing heat accumulation, avoiding laser-plume interaction, and accounting for the direction of the shield gas flow.

Unfortunately, most AM processes require time-consuming and costly post-processing. Examining and optimizing these techniques is an increasingly important research topic to provide methods that reliably produce the required part quality to achieve the industrial acceptance, e.g., laser polishing as a post-processing technique for polyamide 12 samples generated by DED (DED-LB/P/PA12) has been investigated [Kut23]. The effects of laser power and number of runs on surface roughness as well as mechanical properties have been analyzed. The results show that laser polishing can effectively reduce surface roughness and improve the mechanical performance of DED-LB/P/PA12 samples. Welding has also become an important post-process after AM. Brunner–Schwer et al. [Bru22] presented results of welding structures of Inconel 718 produced with PBF-LB/M. No extreme overheating, and melting of flanges of these structures were observed. Despite the local heating of the flange, the cooling rates are notably faster ($-12°C/s$) on the AM side.

In order to reach the global sustainability goals, the knowledge, investigation, and improvement of the ecological footprint of industrial processes become even more relevant. Therefore, calculation methods are developed to gain insight into the value chain of laser processes, e.g., Michel Honoré [Hon23] started asking the question if laser AM is really a green technology. The resulting CO_2 emissions are observed to be strongly dependent on the choice of alloying elements in the alloy as well as on the regional origin of the sourcing of materials and energy. In addition, the alloy chosen in the AM repair affects the lifetime of the component. A few selected scenarios have been already examined, reversing the carbon footprint balance in favor of AM, with an advantage to AM of up to 26% in terms of CO_2 emissions in the most favorable scenario.

3.13.2.2 Fundamentals about Laser Materials Processes

Regarding laser-material interaction fundamentals, high-temperature material data is, in general, difficult to obtain and to find in literature due to the multitude of challenges related to measurements. The measurement of absorption values of liquid steel surfaces at high temperatures is challenging. However, a radiometric measurement setup has enabled the derivation of laser absorption values at high temperatures. The data show a general trend of increasing absorption at increasing temperature with an absorption drop just above boiling temperature [Vol23h]. It is suspected that the vapor above the material's surface absorbs laser energy that reduces the laser energy transferred to the metal's surface. At even higher temperatures, the absorption increases again. However, the measurements still include uncertainties about the actual values of temperature and surface tension at those extreme conditions and require further investigations.

Fundamentals about the interaction between laser light and the metal vapor plume are examined [Wah23]. This phenomenon is, in particular, relevant to laser processes that include the formation of a keyhole such as laser deep penetration welding. The ejected vapor plume from the keyhole opening interacts with the laser beam in terms of scattering, photon absorption, and phase front deformation. A measurement setup has been introduced that helps to characterize the vapor plume. A high-speed camera in combination with spectral measurements has been used to define the interaction locations and the related spectral responses to determine the underlying interaction mechanisms at different distances and regions from the keyhole opening. The thermal emissions and the scattered light of the incident laser beam have been captured. The use of different optical filters allows the distinction between scattered light and thermal emission. The average images of the high-speed camera show zones with different optical emission behavior. This in turn indicates locally different mechanisms of interactions between the vapor plume, the particle cloud, and the incident laser beam. Additionally, spectrometer measurements have confirmed the source of optical emissions. The thermal emission and therefore the temperature of the vapor plume decreases with the increasing distance to the keyhole. Particle formation leads to increased scattering of laser light reaching its maximum in this case at around 100 mm above the keyhole.

3.13.2.3 Monitoring and Control

X-ray observation facilities have become more attractive to observe laser material processes. For laser materials process observations, mainly two monitoring techniques are used, namely diffractometry and radiometry. Diffractometric measurements aim to observe the grain structure inside the material, which can be observed after processing or even during laser interaction [Sil23]. Diffraction rings can be related to the internal structure of the volume illuminated by the X-ray or neutron beam. When recording the diffractometric data during laser impact, the phase transformation kinetics can be observed [Heg21]. This can give indications about the microstructural development during cooling and can help to control and influence the material properties.

Radiometric observations help in visualizing the phenomena during melting and vaporization processes. Current X-ray beams at synchrotrons can penetrate a few millimeters of steel material. By moving the metal sheet during the melting process, the melt pool and vapor channel can be observed. The temporal resolution possible today enables a detailed observation of the melt pool dimensions and flow fields, e.g., Schmidt et al. [Sch23d] used in-situ X-ray monitoring of the laser welding process for analyzing the impact of ambient oxygen on the appearance of the weld seams. Due to the effect of gas-induced dynamic pressure, the shielding of oxygen is also highly relevant to melt pool dynamics and spatter formation due to the growth of oxides and the influence of surface tension. Therefore, the effect of locally supplied argon on oxide growth and seam topography during deep penetration laser beam welding of high-alloy steel AISI 304 has been evaluated. High-speed synchrotron X-ray imaging has been performed to separate the effect of the gas-induced pressure and the gas-induced shielding of a local gas supply on keyhole geometry. The results show that a local supply of argon contributes to a difference in oxide growth, affecting melt pool convection and weld seam geometry. It has been further shown that the effect of local gas flows at low flow rates is primarily because of oxygen shielding, as no significant difference in keyhole geometry has been found by high-speed synchrotron X-ray imaging. Further, Schricker et al. [Sch23c] used a similar X-ray monitoring setup observing the welding of a lap joint of AISI 304 high-alloy steel sheets (X5CrNi18-10/1.4301) with the ability to adjust different gap heights between top and bottom sheet (up to 0.20 mm). A time-resolved description of the "false friend" formation has been provided by visualizing the interaction between keyhole and melt pool during laser welding and solidification processes within the gap area. The bridging of the joining gaps has been limited due to the gap height and insufficient melt supply leading to the solidification of bridge. The distance between the solidified bridge and the keyhole has increased with time, while the keyhole and melt pool dynamics have initiated the formation of new melt bridges whose stability has been defined by melt flow conditions, surface tension, and gap heights.

Vapor channel observations are possible as well for gaining a more detailed understanding of the keyhole processes. Chung et al. [Chu23] demonstrated an X-ray keyhole observation at a frame rate >19 kHz of an aluminum sheet at 0.5 mm thickness and showed the expansion and evolution of the keyhole inside a melt pool, which can enable the more precise calculation of keyhole wall temperatures and pressure balances. In addition, understanding of pore formation is an ongoing topic. In copper hairpin welding for battery applications, high-quality weld seams are, in particular, necessary. X-ray imaging can help in identifying pore formation mechanisms. It has been seen that static beam shaping can reduce the pore volume produced during keyhole welding. When a central spot and a ring spot around are used, a reduction of 60% has been achieved [Oml23]. A reduced dynamic keyhole can be observed when using steel material [Kau23b]. In addition, the newly developed high-power lasers at different wavelengths are tested. At keyhole welding of copper and steel, a tendency to reduce dynamic process behavior has been observed [Kau23b].

Volpp et al. [Vol23g] also used an X-ray observation to observe melt pool impacts on the root instabilities in thick sheet laser beam welding. It has been observed that pore formation at lower laser power occurs at the keyhole tip through keyhole bulging and separation. At increasing laser power, the bulging and separation happen at higher zones of the keyhole. After separation, the gas bubbles remain at the height, where created at low laser power, but are pushed downward in the melt pool at higher laser power. The low resistance of the melt seems to increase the possibility of gas bubbles being pushed downward. Therefore, it is hypothesized that the keyhole bulging can impact the melt flow. This has been shown by relating the bulging events to the slightly delayed root dynamic increase.

The influence of alloying elements on the keyhole stability is also an ongoing research topic, e.g., the impact of magnesium on the melt pool behavior during keyhole welding has been examined, showing that there is a reduced laser beam energy absorbance at a higher content of magnesium in the aluminum alloy [Wag21]. This knowledge can be further developed to provide methods to alter the process behavior using material modifications.

Furthermore, it has been shown that the keyhole geometry also depends on the way it is produced. It has been observed that a keyhole forming from a flat surface can change into deeper, narrower shapes after passing a surface hump without changing any other process parameters [Fau23].

Improved process understanding can be gained by process simulations. Fluid dynamic simulations benefit from more detailed knowledge about material properties at high temperatures and, in return, can provide additional information about melt pool and imperfection mechanisms [Zen23].

3.13.2.4 Process Emission Monitoring

Further details of the relation between process emissions and process behavior are necessary to better understand the process and further developments of optimization routes, e.g., an investigation of the impact of different laser wavelengths on the keyhole during laser processing has been presented by Möbus et al. [Möb23]. Laser beam deep penetration welding experiments have been carried out on 2.4068 pure nickel using an infrared laser source and a blue laser source with comparable beam properties. The experiments have been monitored and compared by a multi-sensor setup and metallographic analyses. This setup includes measurements of airborne acoustic emissions and two high-speed video cameras for spatter tracking and tracking of the keyhole area. The use of a blue laser beam has led to a lower spatter quantity, an increase of porosity, and a significant change in acoustic emissions.

An innovative temperature measurement method has been developed aiming to gain access to temperatures in the vicinity of the keyhole front wall [Por23]. The measurements have been performed at different depths for each combination of process parameters to record an axially resolved temperature profile. Supplementing the temperature measurements with OCT-based keyhole depth measurements and high-speed recordings of the process zone surface has allowed the determination of the time, when the refractory probe has been destroyed, identifying the valid regions of measurements.

Spectral data from the processing zone can provide increased insight into the process. Therefore, line-intensity ratios are under investigation (e.g., [Sch23a]), comparing the line-intensities of substrate and filler-material elements during DED to correlate them with the metallographic results (dilution and chemical composition). An increased degree of dilution leads to a surge in the mixing levels of the substrate and the filler-material within the deposition tracks. Accordingly, line-intensities of elements inside the substrate increase relative to the filler-material elements.

Robust monitoring methods are in development. Track dimension measurements during DED give valuable information that can be used to control process parameters online in the future. A novel in-process monitoring approach for DED-LB/M has been presented with a new developed and robust laser triangulation sensor from Falldorf Sensor GmbH [Sch23b]. It is characterized by having a low height deviation compared to a confocal microscope. The results show that the layer height and width with their peaks and valleys can be monitored even in harsh environments and with a small distance (10 mm) to the process zone. These aspects provide a solid basis for a further control of the layer topology.

Increasing the process stability to prevent spattering and pore formation remains a challenging aspect to be solved in the future.

3.13.2.5   Artificial Intelligence Integration

The potentials of artificial intelligence methods for laser materials processing are an upcoming field of research. Machine learning opportunities are seen in (1) inline monitoring of sensor data and immediate interpretation to enable efficient and reliable process control and (2) parameter prediction based on the desired process results. These methods often require a lot of training data, which is often practically not possible to provide. Therefore, attempts have been made to improve the methods, e.g., a physics-informed hybrid model for the prediction of threshold of deep-penetration laser welding has been presented [Jar23]. A "residual model" approach has been used where a machine learning model is applied to learn and compensate for the deviations of an analytical model to the experimental results. The results show an increase in model accuracy by using such a hybrid model compared to only using the analytical model. In comparison to only using a black-box machine learning model, the amount of required training data can be reduced, and the extrapolation capability can be improved. This indicates that using a hybrid model can significantly reduce the amount of required training data, and hence the number of experiments required to train the model. This is especially beneficial when acquiring the training data is very expensive, as can be the case in laser materials processing.

3.14   SUMMARY AND OUTLOOK

The laser is a powerful and useful tool for laser materials processing, having applications in all production technologies. Laser beam processing is based on local heating by simply transferring the right energy to the right position. This offers achieving a high process flexibility and efficiency. Beam shaping potentials for influencing and stabilizing the processes are not fully explored and will support the processing in the future when better understanding of the impacts will be gained. In addition, monitoring and control aspects will gain importance. Sufficiently convincing process quality data as a measure for the quality of the parts is a mandatory requirement. Monitoring and control tools incorporating machine learning approaches can help in enabling robust and self-adapting processes. Laser processes are increasingly integrated into multi-processing machines to enable several processing steps in one machine for decreasing lead times.

NOTE

1   Courtesy of Fraunhofer IWS and Himani Naesstroem.

4 Symbols and Abbreviations

SYMBOLS

Name	Symbol	Value	Unit
Absorbed laser power	P_A		W
Absorptance	A		
Absorption coefficient	α		1/m
Absorption or extinction index	k		
Absorption efficiency	η_{abs}		
Absorption length	l		m
Ambient temperature	T_{amb}		K
Angles	$\beta_i, \alpha, \theta_b$		rad
Angle of incidence	φ		rad
Beam quality	M^2		
Beam area	A_{beam}		mm
Chemical energy	P_{chem}		W
Convective heat losses	P_{conv}		W
Coordinates	x, y, z		m
Cross-sectional area of the melt pool	A_{MP}		m^2
Refractive index	n		
Diffusion coefficient	D		m^2/s
Diffusion depth	x_d		m
Distance between resonator mirrors	L		m
Divergence angle	θ		mrad
Electrical plug power input	P_{total}		W
Energy difference	ΔE		J
Energy for pumping	E_{pump}		J
Energy	E		J
Focal distance (F…focusing optic, C…collimating optic)	f_D, f_F, f_C		m
Focal beam diameter	d_F		m
F-number	F		
Heat conduction losses	P_{cond}		W
Heat input	Q		W/m
Intensity	I		W/cm^2

DOI: 10.1201/9781003486657-4

Name	Symbol	Value	Unit
Intensity deep penetration threshold	I_{th}		W/cm²
Intensity plasma threshold	I_{pl}		W/cm²
Laser frequency	f		Hz
Laser beam waist radius	w_0		m
Laser beam intensity at interaction	I_0		W/cm²
Laser diameter (on L...lens, C...fiber)	d_L, d_C		m
Laser caustic radius	w		m
Laser power	P_L		W
Latent heat of melting	H_m		kJ7kg
Material density	ρ		kg/m³
Melting temperature	T_m		K
Planck's constant	h	6.625×10^{-34}	J·s
Power	P		W
Power for melting the material volume	P_{th}		W
Process efficiency	$\eta_{process}$		
Process velocity	v		m/min
Quantum efficiency	η_Q		
Radius	r		m
Radius of curvature	R_{gas}		m
Rayleigh length	z_R		m
Reflection losses	P_{refl}		W
Reflectance	R		
Specific heat capacity	c_p		J/(kg*K)
Spherical aberration	Δs		m
Surface energy	γ		N/m
Temperature	T		K
Temperature increase	ΔT_{max}		K
Thermal diffusivity	κ		m²/s
Thermal efficiency	η_{th}		
Thermal focus shift	$\Delta f'$		m
Thermal radiation losses	P_{rad}		W
Time	t		s
Universal gas constant	R		J/(kg*mol)
Vacuum speed of light	c	2.998×10^8	m/s
Vacuum wavelength	λ_0		m
Wall plug efficiency	η_{WP}		
Wavelength of the light	λ		m

ABBREVIATIONS

Al	aluminum
AM	additive manufacturing
BPP	beam parameter product
C	carbon
CaF$_2$	calcium fluoride
CAD	computer aided design
CAM	computer aided manufacturing
CNC	compute numerical control
Co	cobalt
CO$_2$	carbon dioxide
CT	computer tomography
Cu	copper
cw	continuous wave
CYCLAM	reCYCling by laser ablation to feed additive manufacturing
DED	directed energy deposition
DLD	direct laser deposition
DMD	direct metal deposition
DOE	diffractive optical elements
EDX	energy dispersive X-ray analysis
Excime	Excited dimer
FDM	fused deposition modeling
Fe	iron
HAZ	heat-affected zone
He	helium
IR	infrared
KrF	krypton fluoride
LAM	laser active medium
LASER	light amplification by stimulated emission of radiation
LB	laser beam
LMD	laser metal deposition
M	metal
MAG	metal active gas
MAPS	metal additive manufacturing using powder sheets
MASER	microwave amplification by stimulated emission of radiation
MIG	metal inert gas
NDT	non-destructive testing
N$_2$	nitrogen
NA	numerical aperture
Nd	neodymium
Ne	neon
Ni	nickel
OCT	optical coherence tomography
OD	optical density
PBF	powder bed fusion

pw	pulsed width
QCW	quasi-cw
SEM	scanning electron microscope
SiO$_2$	silicon dioxide
SLM	selective laser melting
SLS	selective laser sintering
TEM	transversal electromagnetic mode
TiB$_2$	titanium diboride
TIG	tungsten inert gas
TTT	time-temperature-transformation
UV	ultraviolet
UVFS	ultraviolet degree fused silica (SiO$_2$)
WAAM	wire-arc additive manufacturing
WDX	wavelength dispersive X-ray analysis
XRD	X-ray powder diffraction
YAG	yttrium aluminum garnet
Yb	ytterbium
Zn	zinc
ZnSe	zinc selenide

References

[Ada70] Adams, M. J. (1970). Gas jet laser cutting. In: Needham, J. C. ed. Proceedings of the Conference Advances in Welding Processes, 14–16 April 1970, Harrogate, UK, Abington: The Welding Institute, pp. 140–146.

[All87] von Allmen, M. (1987). Laser-Beam Interactions with Materials, Springer: Berlin.

[Bec96] Beck, M. (1996). Modellierung des Lasertiefschweißens, Dissertation, Universität Stuttgart, B. G. Teubner (Stuttgart).

[Ber07] Bergström, D., Powell, J., & Kaplan, A. F. H. (2007). The absorptance of steels to Nd:YLF and Nd:YAG laser light at room temperature. Applied Surface Science, 253 (11), 5017–5028.

[Ber08] Bergström, D. (2008). The absorption of laser light by rough metal surfaces, Doctoral dissertation, Luleå tekniska universitet.

[Bid18] Bidare, P., Bitharas, I., Ward, R. M., Attallah, M. M., & Moore, A. J. (2018). Fluid and particle dynamics in laser powder bed fusion. Acta Materialia, 142, 107–120.

[Bob80] Bober, M., Singer, J., & Wagner, K. (1980). Spectral reflectivity and emissivity measurements of solid and liquid U02 at 458, 514.5 and 647 nm as a function of polarization and angle of incidence. Seventh Symposium on Thermophysical Properties (Edited by Cezairlyian, A.). pp. 344–350.

[Bol13] Boley, M., Berger, P., Webster, P. J., Weber, R., Van Vlack, C., Fraser, J., & Graf, T. (2013, October). Ivestigating the weld depth behaviour using different observation techniques: X-ray, inline coherent imageing and highspeed observation during welding ice. In International Congress on Applications of Lasers & Electro-Optics (pp. 22–27). AIP Publishing.

[Bro08] Brockmann, R., & Havrilla, D. (2008). Industrial application of high power disk lasers. Proceedings of SPIE - The International Society for Optical Engineering, 6871. 68710I-68710I-8

[Bru17] Brueckner, F., Riede, M., Marquardt, F., Willner, R., Seidel, A., Thieme, S., Leyens, C., & Beyer, E. (2017). Process characteristics in high-precision laser metal deposition using wire and powder. Journal of Laser Applications, 29 (2), 022301.

[Bru22] Brunner-Schwer, C., Simón-Muzás, J., Biegler, M., Hilgenberg, K., & Rethmeier, M. (2022). Laser welding of L-PBF AM components out of Inconel 718. Procedia CIRP, 111, 92–96.

[Bus23] Bussek, T., Völl, A., Stollenwerk, J., Mitri, A., Stollenwerk, A., & Holly, C. (2023). Scan path optimization for laser additive manufacturing with quantum computing. Lasers in Manufacturing Conference, WLT, Munich.

[Cha21] Chang, F. Y., Liang, T. H., Wu, T. J., & Wu, C. H. (2021). Using 3D printing and femtosecond laser micromachining to fabricate biodegradable peripheral vascular stents with high structural uniformity and dimensional precision. The International Journal of Advanced Manufacturing Technology, 116, 1523–1536.

[Che23] Chen, C. W., Nguyen, L. V., Wisal, K. et al. (2023). Mitigating stimulated Brillouin scattering in multimode fibers with focused output via wavefront shaping. Nature Communications, 14, 7343. https://doi.org/10.1038/s41467-023-42806-1

[Cho20a] Cho, W. I., & Woizeschke, P. (2020). Analysis of molten pool behavior with buttonhole formation in laser keyhole welding of sheet metal. International Journal of Heat and Mass Transfer, 152, 119528.

[Cho20b] Choi, J., & Schwarz, C. (2020). Advances in femtosecond laser processing of optical material for device applications. International Journal of Applied Glass Science, 11 (3), 480–490.

[Chu23] Chung, W. S., Hummel, M., Spurk, C., Häusler, A., Olowinsky, A., Häfner, C., ... & Moosmann, J. (2023). In situ X-ray phase contrast imaging of the melt and vapor capillary behavior during the welding regime transition on aluminum with limited material thickness. Welding in the World, 68 (1), 43–50.

[Dau93] Dausinger, F., & Shen, J. (1993). Energy coupling efficiency in laser surface treatment. ISIJ International, 33 (9), 925–933.

[Dau95] Dausinger F. (1995). Strahlwerkzeug Laser: Energieeinkopplung und Prozeßeffektivität. Universität Stuttgart, Habilitationsschrift. – Laser in der Materialbearbeitung: Forschungsberichte des IFSW, B.G. Teubner Stuttgart.

[Dem18] Demir, A. G. (2018). Micro laser metal wire deposition for additive manufacturing of thin-walled structures. Optics and Lasers in Engineering, 100, 9–17.

[Dew20a] Dewi, H. S., & Volpp, J. (2020). Impact of laser beam oscillation strategies on surface treatment of microalloyed steel. Journal of Laser Applications, 32 (4), 042006.

[Dew20b] Dewi, H. S., Fischer, A., Volpp, J., Niendorf, T., & Kaplan, A. F. (2020). Microstructure and mechanical properties of laser surface treated 44MnSiVS6 microalloyed steel. Optics & Laser Technology, 127, 106139.

[Dew20c] Dewi, H. S., Volpp, J., & Kaplan, A. F. (2020). Short thermal cycle treatment with laser of vanadium microalloyed steels. Journal of Manufacturing Processes, 57, 543–551.

[Dew21] Dewi, H. S., Volpp, J., Frostevarg, J., & Siltanen, J. (2021, November). Influence of mill scale on oxygen laser cutting processes. In IOP Conference Series: Materials Science and Engineering (Vol. 1135, No. 1, p. 012008), IOP Publishing: Bristol, England.

[Dew22] Dewi, H. S., Volpp, J., & Kaplan, A. F. (2022). Influence of secondary-pass laser treatment on retained ferrite and martensite in 44MnSiVS6 microalloyed steel. Materials Today Communications, 31, 103282.

[Dre02] Drelich, J., Fang, C., & White, C. L. (2002). Measurement of interfacial tension in fluid-fluid systems. The Encyclopedia of Surface and Colloid Science, 3, 3158–3163.

[Dru00] Drude, P. (1900). Zur Elektronentheorie der Metalle. Annalen der Physik, 566, 1.

[Dul99] Duley, W. W. (1999). Laser Welding, John Wiley and Sons: New York.

[Ein05] Einstein, A. (1905). On the electrodynamics of moving bodies. Annalen der Physik, 17, 891–921. (In German.)

[Eus13] Eustathopoulos, N., Hodaj, F., & Kozlova, O. (2013). The wetting process in brazing. In Advances in Brazing (pp. 3–30), Woodhead Publishing: Sawston, Cambridge, UK.

[Fab05] Fabbro, R., Slimani, S., Coste, F., & Briand, F. (2005). Study of keyhole behaviour for full penetration Nd–Yag CW laser welding. Journal of Physics D: Applied Physics, 38, 1881–1887.

[Fab10] Fabbro, R. (2010). Melt pool and keyhole behaviour analysis for deep penetration laser welding. Journal of Physics D: Applied Physics, 43 (44), 445501.

[Fau23] Faue, P., Rathmann, L., Möller, M., Hassan, M., Clark, S. J., Fezzaa, K., Klingbeil, K., Richter, B., Volpp, J., Radel, T., & Pfefferkorn, F. E. (2023). High-speed X-ray study of process dynamics caused by surface features during continuous-wave laser polishing. CIRP Annals – Manufacturing Technology, 72 (1), 201–204.

[Fed20] Fedina, T., Sundqvist, J., Powell, J., & Kaplan, A. F. (2020). A comparative study of water and gas atomized low alloy steel powders for additive manufacturing. Additive Manufacturing, 36, 101675.

[Fin90] Finke, B. R., Finke, M., Kapadia, P. D., Dowden, J. M., & Simon, G. (1990, August). Numerical investigation of the Knudsen-layer, appearing in the laser-induced evaporation of metals. In Laser-Assisted Processing II (Vol. 1279, pp. 127–134), SPIE.

[Fri00] Friedrich, R., Radons, G., Ditzinger, T., & Henning A. (2000). Ripple formation through an interface instability from moving growth and erosion sources. Physical Review Letters, 85 (23), 4884–4887.

[Gar19] García-Sesma, L., López, B., & Pereda B. (2019). Effect of coiling conditions on the strengthening mechanisms of Nb microalloyed steels with high Ti addition levels. Materials Science and Engineering: A, 748, 386–395. https://doi.org/10.1016/j.msea.2019.01.105

[Gas93] Gasser, A. (1993). Oberflächenbehandlung metallischer Werkstoffe mit CO_2-Laserstrahlung in der flüssigen Phase, Dissertation, RWTH Aachen.

[Gat12] Gatzen, M. (2012). Influence of low-frequency magnetic fields during laser beam welding of aluminium with filler wire. Physics Procedia, 39, 59–66.

[Gat14] Gatzen, M., Radel, T., Thomy, C., & Vollertsen, F. (2014). Wetting behavior of eutectic Al–Si droplets on zinc coated steel substrates. Journal of Materials Processing Technology, 214 (1), 123–131.

[Gie05] Giesen, A., & Dausinger, F. (2005). Thin-disk solid-state lasers and their applications. The Review of Laser Engineering, 33 (4), 219–222.

[Gon07] Gonzalez, J. J., Freton, P., & Masqu`ere, M. (2007). Experimental quantification in thermal plasma medium of the heat flux transferred to an anode material. Journal of Physics D, 40 (18), 5602–5611.

[Gon23] Gonzalez-Hernandez, D., Varapnickas, S., Bertoncini, A., Liberale, C., & Malinauskas, M. (2023). Micro-optics 3D printed via multi-photon laser lithography. Advanced Optical Materials, 11 (1), 2201701.

[Gug23] Mi, Y., Guglielmi, P., Nilsen, M., Sikström, F., Palumbo, G., & Ancona, A. (2023). Beam shaping with a deformable mirror for gap bridging in autogenous laser butt welding. Optics and Lasers in Engineering, 169, 107724.

[Har95] Hart, P. H. M., & Mitchell, P. S. (1995). The effect of vanadium on the toughness of welds in structural and pipeline steels. Welding Journal, 74, 239s–248s.

[Hau21] Hauser, T., Reisch, R. T., Breese, P. P., Nalam, Y., Joshi, K. S., Bela, K., … & Kaplan, A. F. (2021). Oxidation in wire arc additive manufacturing of aluminium alloys. Additive Manufacturing, 41, 101958.

[Hau22] Hauser, T., Reisch, R. T., Kamps, T., Kaplan, A. F., & Volpp, J. (2022). Acoustic emissions in directed energy deposition processes. The International Journal of Advanced Manufacturing Technology, 119, 3517–3532.

[Heg21] Hegele, P., von Kobylinski, J., Hitzler, L., Krempaszky, C., & Werner, E. (2021). In-situ XRD study of phase transformation kinetics in a Co-Cr-W-alloy manufactured by laser powder-bed fusion. Crystals, 11 (2), 176.

[Hei13] Heider, A., Sollinger, J., Abt, F., Boley, M., Weber, R., & Graf, T. (2013). High-speed X-ray analysis of spatter formation in laser welding of copper. Physics Procedia, 41, 112–118.

[Het76] Hettche, L. R., Tucker, T. R., Schriempf, J. T., Stegman, R. L., & Metz, S. A. (1976). Mechanical response and thermal coupling of metallic targets to high-intensity 1.06-µ laser radiation. Journal of Applied Physics, 47 (4), 1415–1421.

[Hon23] Honoré, M., & Hansen, S. K. (2023). Is AM always the green manufacturing alternative? A comparative study of carbon footprint. Lasers in Manufacturing Conference, WLT, Munich.

[Hua18] Huang, L., Hua, X., Wu, D., Fang, L., Cai, Y., & Ye, Y. (2018). Effect of magnesium content on keyhole-induced porosity formation and distribution in aluminum alloys laser welding. Journal of Manufacturing Processes, 33, 43–53.

[Hul57] van de Hulst, H. C. (1957). Light Scattering by Small Particles, Dover Publications Inc.: New York.

[Ind18] Indhu, R., Vivek, V., Sarathkumar, L., Bharatish, A., & Soundarapandian, S. (2018). Overview of laser absorptivity measurement techniques for material processing. Lasers in Manufacturing and Materials Processing, 5 (4), 458–481.

[Jar23] Jarwitz, M., & Michalowski, A. (2023). Hybrid model for the threshold of deep-penetration laser welding. Lasers in Manufacturing Conference, WLT, Munich.

[Jay24] Devara, J. S., Jakhar, S., Sihag, Y., Panda, B., Venkatesan, A., & Singh, K. P. (2024). Development of an all-solid-state air-cooled high-power blue diode laser for metal processing. Optics Letters, 49, 17–20.

[Jia19] Jiang, M., Chen, X., Chen, Y., & Tao, W. (2019). Increasing keyhole stability of fiber laser welding under reduced ambient pressure. Journal of Materials Processing Technology, 268, 213–222.

[Kap13] Eriksson, I., Powell, J., & Kaplan, A. F. (2013). Melt behavior on the keyhole front during high speed laser welding. Optics and Lasers in Engineering, 51 (6), 735–740.

[Kar10] Karunakaran, K. P., Suryakumar, S., Pushpa, V., & Akula, S. (2010). Low cost integration of additive and subtractive processes for hybrid layered manufacturing. Robotics and Computer-Integrated Manufacturing, 26 (5), 490–499.

[Kat02] Katayama, S., & Matsunawa, A. (2002). Microfocused X-ray transmission real-time observation of laser welding phenomena. Welding International, 16 (6), 425–431.

[Kau23] Kaufmann, F., Forster, C., Hummel, M., Olowinsky, A., Beckmann, F., Moosmann, J., … & Schmidt, M. (2023). Characterization of vapor capillary geometry in laser beam welding of copper with 515 nm and 1030 nm laser beam sources by means of in situ synchrotron X-ray imaging. Metals, 13 (1), 135.

[Kau23b] Kaufmann, F., Schrauder, J., Hummel, M., Spurk, C., Olowinsky, A., Beckmann, F., … & Schmidt, M. (2023). Towards an understanding of the challenges in laser beam welding of copper–observation of the laser-matter interaction zone in laser beam welding of copper and steel using in situ synchrotron X-ray imaging. Lasers in Manufacturing and Materials Processing, 11, 37–76.

[Kaw18] Kawahito, Y., Wang, H., Katayama, S., & Sumimori, D. (2018). Ultra high power (100 kW) fiber laser welding of steel. Optics Letters, 43 (19), 4667–4670.

[Kaz09] Kazemi, K., & Goldak, J. A. (2009). Numerical simulation of laser full penetration welding. Computational Materials Science, 44 (3), 841–849.

[Kee88] Keene, B. J. (1988). Review of data for the surface tension of iron and its binary alloys. International Materials Reviews, 33 (1), 1–37.

[Ken23] Kengne, E., & Lakhssassi, A. (2023). Femtosecond solitons and double-kink solitons in passively mode-locked lasers. Optical and Quantum Electronics, 55 (6), 565.

[Kho22] Khorasani, M., Ghasemi, A., Leary, M., Sharabian, E., Cordova, L., Gibson, I., … & Rolfe, B. (2022). The effect of absorption ratio on meltpool features in laser-based powder bed fusion of IN718. Optics & Laser Technology, 153, 108263.

[Ki02a] Ki, H., Mazumder, J., & Mohanty, P. S. (2002). Modeling of laser keyhole welding: Part I. Mathematical modeling, numerical methodology, role of recoil pressure, multiple reflections, and free surface evolution. Metallurgical and Materials Transactions A, 33 (6), 1817–1830.

[Ki02b] Ki, H., Mazumder, J., & Mohanty, P. S. (2002). Modeling of laser keyhole welding: Part II. Simulation of keyhole evolution, velocity, temperature profile, and experimental verification. Metallurgical and Materials Transactions A, 33 (6), 1831–1842.

[Kla84] Klauminzer, G. K. (1984). Twenty years of commercial laser s – A capsule history. Laser Focus/Electro-Optics, 20, (December), 54–79.

[Kog20] Kogel-Hollacher, M., Strebel, M., Staudenmaier, C., Schneider, H. I., & Regulin, D. (2020, March). OCT sensor for layer height control in DED using SINUMERIK® controller. In Laser 3D Manufacturing VII (Vol. 11271, pp. 59–63), SPIE: Bellingham, Washington.

[Küg19] Kügler, H., & Vollertsen, F. (2019). Influences of surface pretreatments on absorptivity changes induced by laser beam processing. In: Proceedings of LAMP2019 - the 8th International Congress on Laser Advanced Materials Processing.

[Kut23] Kutlu, Y., Schuleit, M., Thiele, M., Esen, C., & Ostendorf, A. (2023). Laser polishing as a post-processing tool for DED-LB/P. Lasers in Manufacturing Conference, WLT, Munich.

[Las14] Laskin, A., & Laskin, V. (2014). Freeform beam shaping for high-power multimode lasers. In: Proceedings of SPIE - The International Society for Optical Engineering; 89601P-1 – 89601P-12.

[Las18] Laserfocus. (2018). Annual laser market review & forecast: Lasers enabling lasers [Internet]. Available from: https://www.laserfocusworld.com/articles/print/volume-54/issue-01/features/annual-laser-market-review-forecast-lasers-enabling-lasers.html [Accessed: 2019-04-20]

[Las21a] Laskin, A., & Volpp, J. (2021). Comparison of the thermal focus shift and aberration between the single-mode and multimode lasers. Journal of Laser Applications, 33 (4), 042026.

[Las21b] Laskin, A., Volpp, J., Laskin, V., Nara, T., & Jung, S. R. (2021). Multispot optics for beam shaping of high-power single-mode and multimode lasers. Journal of Laser Applications, 33 (4), 042046.

[Lei21] Leis, A., Bechler, S., Weber, R., & Graf, T. (2021). Laser-based powder bed fusion with 16 kW. In Lasers in Manufacturing Conference 2021, Munich, Germany, WLT – Wissenschaftliche Gesellschaft Lasertechnik e.V., Contribution 243.

[Lin01] Lindenau, D., Ambrosy, G., Berger, P., & Hügel, H. (2001, October). Effects of magnetically supported laser beam welding of aluminium alloys. In International Congress on Applications of Lasers & Electro-Optics (Vol. 2001, No. 1, pp. 168–178), Laser Institute of America: Orlando, FL.

[Liu04] Liu, X., & Cheng, C. H. (Apr. 2004). System and method of laser drilling. U.S. Patent No. 6,720,519. 13.

[Lup22] Lupoi, R., Abbott, W. M., Senthamaraikannan, R., McConnell, S., Connolly, J., Yin, S., et al. (2022). Metal additive manufacturing via a novel composite material using powder and polymers formed in sheets. CIRP Annals.

[Mai60] Maiman, T. H. (1960). Pulsed ruby laser, USA; Optical and microwave-optical experiments in ruby. Physical Review Letters, 4 (11), 564–566.

[Mak92] Makashev, N. K., Asmolov, N. S., Blinkov, V. V., Boris, A. Y., Buzykin, O. G., Burmistrov, A. V., Gryaznov, M. R., & Makarov, V. A. (1992). Gas hydrodynamics of metal cutting by CW laser radiation in a rare gas. Soviet Journal of Quantum Electronics, 22, 847–852.

[Mar11] Marchand, A., Weijs, J. H., Snoeijer, J. H., & Andreotti, B. (2011). Why is surface tension a force parallel to the interface? American Journal of Physics, 79 (10), 999–1008.

[Mat10] https://www.utwente.nl/en/et/ms3/research-chairs/lp/Matlab%20Laser%20Toolbox/

[Mat05] Matsumoto, T., Fujii, H., Ueda, T., Kamai, M., Nogi, K. (2005). Measurement of surface tension of molten copper using the free-fall oscillating drop method. Measurement Science and Technology, 16 (2), 432.

[Mis18] Mishra, P., Ilar, T., Brueckner, F., & Kaplan, A. F. H. (2018). Energy efficiency contributions and losses during selective laser melting. Journal of Laser Applications, 30 (3), 032304.

[Miy03] Miyamoto, I. (2003). Precision microwelding with Yb single-mode fiber laser. In: Proceedings of 22th ICALEO 2003, Laser Microfabrication Conference.

[Möb23] Möbus, M., Pordzik, R., Krämer, A., & Mattulat, T. (2023). Process comparison of laser deep penetration welding in pure nickel using blue and infrared wavelengths. IIW annual assembly, Singapore. Doc.IV-1563-2023

[Mod03] Modest, M. (2003). Radiative Heat Transfer, 2nd Ed., Elsevier Science: Amsterdam, The Netherlands.

[Möl16] Möller, F. (2016). Wechselwirkung Zwischen Lichtbogen Und Laserstrahl Beim Fügen Von Aluminium, BIAS Verlag: Bremen, Germany.

[Mor11] Morohoshi, K., Uchikoshi, M., Isshiki, M., Fukuyama, H. (2011). Surface tension of liquid iron as functions of oxygen activity and temperature. ISIJ International, 51 (10), 1580–1586.

[Nel22] Nellian, A. S., & Pang, J. H. L. (2022). Metal additive manufacturing repair study on rail steel with Stellite 6 powder. Materials Today: Proceedings, 70, 101–105.

[Nem97] Nemchinsky, V. A. (1997). Dross formation and heat transfer during plasma arc cutting. Journal of Physics D: Applied Physics, 30, 2566–2572.

[Now15] Nowotny, S., Brueckner, F., Thieme, S., Leyens, C., & Beyer, E. (2015). High-performance laser cladding with combined energy sources. Journal of Laser Applications, 27 (S1), S17001.

[Oli16] Oliveira, J. P., Fernandes, F. B., Miranda, R. M., Schell, N., & Ocaña, J. L. (2016). Residual stress analysis in laser welded NiTi sheets using synchrotron X-ray diffraction. Materials & Design, 100, 180–187.

[Ols16] Reisgen, U., Olschok, S., Jakobs, S., & Turner, C. (2016). Laser beam welding under vacuum of high grade materials. Welding in the World, 60, 403–413.

[Oml23] Omlor, M., Reinheimer, E. N., Butzmann, T., & Dilger, K. (2023). Investigations on the formation of pores during laser beam welding of hairpin windings using a high-speed x-ray imaging system. Journal of Laser Applications, 35 (3), 032010.

[Ott11] Otto, A., Koch, H., Leitz, K. H., & Schmidt, M. (2011). Numerical simulations-a versatile approach for better understanding dynamics in laser material processing. Physics Procedia, 12, 11–20.

[Par07] Partes, K., Habedank, G., Seefeld, T., & Vollertsen, F. (2007, October). Laser alloying of aluminum using a deep penetration process with fiber laser. In International Congress on Applications of Lasers & Electro-Optics (Vol. 2007, No. 1, p. 1606), Laser Institute of America: Orlando, FL.

[Pep11] Pépe, N., Egerland, S., Colegrove, P. A., Yapp, D., Leonhartsberger, A., & Scotti, A. (2011). Measuring the process efficiency of controlled gas metal arc welding processes. Science and Technology of Welding and Joining, 16 (5), 412–417.

[Poc17] Pocorni, J., Han, S. W., Cheon, J., Na, S. J., Kaplan, A. F., & Bang, H. S. (2017). Numerical simulation of laser ablation driven melt waves. Journal of Manufacturing Processes, 30, 303–312.

[Pok63] Pokrovskii, N. L., Pugachevich, P. P., & Golubev, N. A. (1963) The role of surface phenomena in metallurgy. Russian Journal of Physical Chemistry A, 43 (8), 1212.

[Pop11] Poprawe, R., Boucke, K., & Hoffman, D. (2011). Tailored Light (p. 2), Springer: Berlin.

[Por23] Pordzik, R., Mattulat, T., & Woizeschke, P. (2023). Frontal pyrometric snapshot measurements of the keyhole wall temperature in laser welding of pure aluminium. Welding in the World, 68 (1), 117–136.

[Pra18] Prasad, H. S., Brueckner, F., & Kaplan, A. F. (2018, October). Powder catchment in laser metal deposition. In International Congress on Applications of Lasers & Electro-Optics, AIP Publishing: College Park, MD.

[Pra20] Siva Prasad, H., Brueckner, F., Volpp, J., & Kaplan, A. F. (2020). Laser metal deposition of copper on diverse metals using green laser sources. The International Journal of Advanced Manufacturing Technology, 107, 1559–1568.

[Pri95] Price, D. F., More, R. M., Walling, R. S., Guethlein, G., Shepherd, R. L., Stewart, R. E., & White, W. E. (1995). Absorption of ultrashort laser pulses by solid targets heated rapidly to temperatures 1–1000 eV. Physical Review Letters, 75 (2), 252.

[Pri20] Prieto, C., Vaamonde, E., Diego-Vallejo, D., Jimenez, J., Urbach, B., Vidne, Y., & Shekel, E. (2020). Dynamic laser beam shaping for laser aluminium welding in e-mobility applications. Procedia CIRP, 94, 596–600.

[Pri23] Primus, T., Novák, M., Zeman, P., & Holešovský, F. (2023). Femtosecond laser processing of advanced technical materials.

[Rat22] Rathmann, L., & Radel, T. (2022). On the use of LIPSS in single-and multi-scale laser-structured tool surfaces under lubricated conditions. Journal of Manufacturing Processes, 77, 819–830.

[Rei21] Reinheimer, E., Hummel, M., Olowinsky, A., Weber, R., & Graf, T. (2021). High-speed synchrotron x-ray imaging of the formation of wedge-shaped capillaries during laser-beam welding at high feed rates. In: Proceedings of the Lasers in Manufacturing Conference.

[Rei23] Reich, S., Goesmann, M., Heunoske, D., Schäffer, S., Lueck, M., Wickert, M., & Osterholz, J. (2023). Change of dominant material properties in laser perforation process with high-energy lasers up to 120 kilowatt. Scientific Reports, 13 (1), 21611.

[Röm10] Römer, G. W., & Huis in 't Veld, A. J. (2010, September). Matlab laser toolbox. In International Congress on Applications of Lasers & Electro-Optics (Vol. 2010, No. 1, pp. 523–529), Laser Institute of America: Orlando, FL.

[Sam17] Samarjy, R. S. M., & Kaplan, A. F. H. (2017). Using laser cutting as a source of molten droplets for additive manufacturing: A new recycling technique. Materials and Design, 125, 76–84.

[Sch87] Schulz, W., et al. (1987). On laser fusion cutting of metals. Journal of Physics D: Applied Physics, 20, 481.

[Sch10] Schaaf, P. (2010). Laser Processing of Materials, Springer-Verlag: New York. https://doi.org/10.1007/978-3-642-13281-0

[Sch19] Shmyrov, A., Mizev, A., Shmyrova, A., & Mizeva, I. (2019). Capillary wave method: An alternative approach to wave excitation and to wave profile reconstruction. Phys Fluids, 31 (1), 012101.

[Sch23a] Schmidt, M., Partes, K., Rajput, R., Phochkhua, G., & Köhler, H. (2023). Dilution monitoring using inline optical emission spectroscopy during Directed Energy Deposition process of aluminium bronze. Lasers in Manufacturing Conference, WLT, Munich.

[Sch23b] Schinderling, A., Bohlen, A., & Seefeld, T. (2023). In-process monitoring and measurement of track geometry for laser metal deposition with laser triangulation. Lasers in Manufacturing Conference, WLT, Munich.

[Sch23c] Schricker, K., Diegel, C., Schmidt, L., Seibold, M., Friedmann, H., Fröhlich, F., … & Bergmann, J. P. (2023). Understanding the formation of "false friends"(hidden lack of fusion defects) in laser beam welding by means of high-speed synchrotron X-ray imaging. Welding in the World, 67 (11), 2557–2570.

[Sch23d] Schmidt, L., Schricker, K., Diegel, C., Sachs, F., Bergmann, J. P., Knauer, A., Romanus, H., Requardt, H., Chen, Y., & Rack, A. (2023). Effect of local and global gas atmosphere on surface-driven phenomena in deep penetration laser beam welding with high welding speeds. IIW annual assembly, Singapore.

[Sem06] Semak, V. V., Knorovsky, G. A., MacCallum, D. O., & Roach, R. A. (2006) Effect of surface tension on melt pool dynamics during laser pulse interaction. Journal of Physics D: Applied Physics, 39 (3), 590.

[Sey99] Seyhan, I., & Egry, I. (1999). The surface tension of undercooled binary iron and nickel alloys and the effect of oxygen on the surface tension of Fe and Ni. International Journal of Thermophysics, 20, 1017–1028.

[Sil23] Silveira, A. C. D. F., Fechte-Heinen, R., & Epp, J. (2023). Microstructure evolution during laser-directed energy deposition of tool steel by in situ synchrotron X-ray diffraction. Additive Manufacturing, 63, 103408.

[Spi69] Spisz, E. W., Weigund, A. J., Bowmun, R. L., & Juck, J. R. (1969). Solar absorptances and spectral reflectances of 12 metals for temperatures ranging from 300 to 500 K. NASA, Tn D-5353.

[Sun24] Sun, D. R., Wang, G., Li, Y., Yu, Y., Shen, C., & Wang, Y. (2024). Laser drilling in silicon carbide and silicon carbide matrix composites. Optics & Laser Technology, 170, 110166.

[Tan20] Tang, C., Le, K. Q., & Wong, C. H. (2020). Physics of humping formation in laser powder bed fusion. International Journal of Heat and Mass Transfer, 149, 119172.

[Tho82] Thomas, T. (1982). Rough Surfaces, Longman: New York.

[Tho20] Thomsen, A. N. (2020). Laser Forming of Sheet metal, PhD thesis, University Aalborg, DK.

[Tri88] Trinh, E. H., Marston, P. L., & Robey, J. L. (1988). Acoustic measurement of the surface tension of levitated drops. Journal of Colloid and Interface Science, 124 (1), 95–103.

[Vic87] Vicanek, M. et al. (1987). Hydrodynamical instability of melt flow in laser cutting. Journal of Physics D: Applied Physics, 20, 140.

[Vol14] Volpp, J., Dietz, T., & Vollertsen, F. (2014). Particle property impact on its distribution during laser deep alloying processes. Physics Procedia, 56, 1094–1101.

[Vol16a] Volpp, J., & Vollertsen, F. (2016). Keyhole stability during laser welding—Part I: Modeling and evaluation. Production Engineering, 10 (4), 443–457.

[Vol16b] Volpp, J., Srowig, J., & Vollertsen, F. (2016). Spatters during laser deep penetration welding with a bifocal optic. In Advanced Materials Research (Vol. 1140, pp. 123–129), Trans Tech Publications: Zurich, Switzerland.

[Vol17] Volpp, J. (2017a). Keyhole stability during laser welding—Part II: Process pores and spatters. Production Engineering, 11 (1), 9–18.

[Vol18a] Volpp, J. (2018). Formation mechanisms of pores and spatters during laser deep penetration welding. Journal of Laser Applications, 30 (1), 012002.

[Vol18b] Volpp, J., Prasad, H. S., Riede, M., Brueckner, F., & Kaplan, A. F. H. (2018). Powder particle attachment mechanisms onto liquid material. Procedia CIRP, 74, 140–143.

[Vol19a] Volpp, J., & Vollertsen, F. (2019). Impact of multi-focus beam shaping on the process stability. Optics & Laser Technology, 112, 278–283.

[Vol19b] Volpp, J., Brueckner, F., & Kaplan, A. F. H. (2019). Track geometry variations in selective laser melting processes. Journal of Laser Applications, 31 (2), 022310.

[Vol21a] Volpp, J., & Laskin, A. (2021). Beam shaping solutions for stable laser welding: Multifocal and multispot beams to bridge gaps and reduce spattering. Photonics Views, 18 (5), 38–41.

[Vol21b] Volpp, J., & Frostevarg, J. (2021). Elongated cavities during keyhole laser welding. Materials & Design, 206, 109835.

[Vol23a] Volpp, J., Naesstroem, H., Wockenfuss, L., Schmidt, M., & Partes, K. (2023). Spectral visualization of alloy reactions during laser melting. Alloys, 2 (3), 140–147.

[Vol23b] Volpp, J. (2023). Impact of melt flow and surface tension on gap bridging during laser beam welding.

[Vol23c] Volpp, J. (2023). Laser light absorption and Brewster angle on liquid metal. Journal of Applied Physics, 133 (20), 205902.

[Vol23d] Volpp, J. (2023). Laser beam absorption measurement at molten metal surfaces. Measurement, 209, 112524.

[Vol23e] Volpp, J. (2023). Surface tension of steel at high temperatures. SN Applied Sciences, 5 (9), 237.

[Vol23f] Volpp, J., Zhang, W., Abbott, W., Coban, A., Casati, R., Padamati, R., Lupoi, R., Marola, S. (2023). Binder Evaporation During Powder Sheet Additive Manufacturing. In: Beaman, J. ed. Proceedings of 2023 Annual International Solid Freeform Fabrication Symposium (SFF Symp 2023).

[Vol23g] Volpp, J., Sato, Y., Tsukamoto, M. (2023). Keyhole instabilities in thick sheet laser beam welding. IIW annual assembly, Singapore. Doc.IV-1556-2023.

[Vol23h] Volpp, J. (2023). High-temperature laser absorption of steel. Lasers in Manufacturing Conference, WLT, Munich.

[Wag21] Wagner, J., Hagenlocher, C., Hummel, M., Olowinsky, A., Weber, R., & Graf, T. (2021). Synchrotron X-ray analysis of the influence of the magnesium content on the absorptance during full-penetration laser welding of aluminum. Metals, 11 (5), 797.

[Wah23] Wahl, J., Frey, C., Sawannia, M., Olschok, S., Weber, R., Hagenlocher, C., Michalowski, A., & Graf, T. (2023). Measurement of the influence of the vapor plume on laser beam characteristics during laser beam welding. Lasers in Manufacturing Conference, WLT, Munich.

[Wan22] Wang, L., Gao, X., & Kong, F. (2022). Keyhole dynamic status and spatter behavior during welding of stainless steel with adjustable-ring mode laser beam. Journal of Manufacturing Processes, 74, 201–219.

[Web10] Weberpals, J.-P. (2010). Nutzen und Grenzen guter Fokussierbarkeit beim Laserschweißen. In: Laser in der Materialbearbeitung, Forschungsberichte der IFSW, T. Graf (Hrsg.), Herbert Utz Verlag.

[Web14] Webster, P. J., Wright, L. G., Ji, Y., Galbraith, C. M., Kinross, A. W., Van Vlack, C., & Fraser, J. M. (2014). Automatic laser welding and milling with in situ inline coherent imaging. Optics Letters, 39 (21), 6217–6220.

[Win17] Winterkorn, R., Petrat, T., Gumenyuk, A., & Rethmeier, M. (25 June 2017). Temperature generation of different travel path strategies to build layers using laser metal deposition. In: Proceedings of Laser in Manufacturing Conference (LiM), Munich, Germany, (WLT e.G., Hannover, Germany), contribution 103.

[Wis85] Wissenbach, K. (1985). Umwandlungshärten mit CO2-Laserstrahlung, Dissertation, TH Darmstadt.

[Xu17] Xu, X., Mi, G., Wang, C. (2017). Laser metal deposition with 316L stainless wire. In: Proceedings of the 37th International Congress on Applications of Lasers and Electro-Optics (ICALEO), LIA Congress Proceedings, Orlando, FL: Laser Institute of America; paper 502.

[Yad10] Yadroitsev, I., Gusarov, A., Yadroitsava, I., & Smurov, I. (2010). Single track formation in selective laser melting of metal powders. Journal of Materials Processing Technology, 210 (12), 1624–1631.

[Yad11] Yadroitsev, I., & Smurov, I. (2011). Surface morphology in selective laser melting of metal powders. Physics Procedia, 12, 264–270.

[Zan23] Zang, T., Wang, Z., Chen, L., Kong, M., Gao, S., Ngwangwa, H. M., … & Zheng, H. (2023). Influence of pulse energy on surface integrity of AZ31 magnesium alloy processed by femtosecond laser shock peening. Journal of Materials Research and Technology, 25, 4425–4440.

[Zel66] Zeldovich, Y. B., & Raizer, Y. P. (1966). Physics of Shock Waves and High-Temperature Hydrodynamic Phenomena I, Academic Press: New York.

[Zen23] Zenz, C., Buttazzoni, M., Ceniceros, M. M., Vázquez, R. G., Puchades, J. R. B., Griñán, L. P., & Otto, A. (2023). Simulation-based process optimization of laser-based powder bed fusion by means of beam shaping. Additive Manufacturing, 77, 103793.

[Zha10] Zhang, Y. Z., Meacock, C., & Vilar, R. (2010). Laser powder micro-deposition of compositional gradient Ti–Cr alloy. Materials & Design, 31 (8), 3891–3895.

[Zha23a] Zhang, N., Wang, M., Ban, M., Guo, L., & Liu, W. (2023). Femtosecond laser drilling 100 µm diameter micro holes with aspect ratios> 20 in a nickel based superalloy. Journal of Materials Research and Technology, 28, 1415–1422.

[Zha23b] Zhang, L., Zhai, W., Bi, G., Xu, S., Lu, G., & Zhou, W. (July 2023). Bonding strengthening through oxidation inhibition by adding TiC particles in laser directed energy deposition of 316L stainless steel. Proc. of the 76th IIW Annual Assembly and Intl. Conf. on Welding and Joining (IIW 2023) 16–21, Singapore. Wei, Z.& Pang, J. eds. ISBN: 978-981-18-7859-6.

[Zha23c] Zhang, T., Guo, B., Jiang, L., Chen, M., & Zhan, N. (2023). Ultrafast observation of multiple shock waves evolution and interaction processes in femtosecond laser processing. Physics of Fluids, 35 (5), 057114.

[Zho15] Zhong, C., Biermann, T., Gasser, A., & Poprawe, R. (2015). Experimental study of effects of main process parameters on porosity, track geometry, deposition rate, and powder efficiency for high deposition rate laser metal deposition. Journal of Laser Applications, 27 (4), 042003.

[Zhu04] Zhu, Q. (2004). Modeling and measurements of the bidirectional reflectance of microrough silicon surfaces, PhD thesis, Georgia Institute of Technology.

Index

3D printing 119

A

absorption 2, 27, 36, 71, 75, 76, 125, 134
absorption length 37
additive manufacturing 119
alloying 51, 115

B

beam caustic 21, 29
beam guiding 22, 26, 117, 125
beam measurement 21, 30
beam parameter product 7, 19, 29
beam quality parameter 20, 28
beam radius 19, 29
beam shaping 22, 30, 54, 59, 74, 97
beam splitter 33
beam waist 7, 19, 30, 98
bending 61
Boltzmann equilibrium 4
brazing 51, 64
Brewster angle 38
buttonhole 99

C

camera 49, 138, 145
caustic 21, 29, 50
cladding 51, 116, 126
CO_2 laser 9
coatings 25, 26
coherence 19
collimator 29
convection 67, 75
cooling 3, 44, 56, 104
cracks 103, 131
cross jet 81
cutting 106

D

deep penetration welding 66, 71
deformation 14, 45, 62, 75, 94
denudation 136, 138
diffusion 45, 81
digital twin 122, 133
dilution 117

diode lasers 17
directed energy deposition (DED) 123, 148
disk lasers 14
dispersing 51, 115
divergence 19, 28, 29
donut beam 33
drilling 111
dross 110, 113
Drude theory 38
dynamics 74, 135

E

electric glow discharge 9, 11
electromagnetic wave 37
emissions 46
energy coupling 16
energy losses 8, 74, 75
excitation 9

F

F-number 28, 29
fiber 23, 24, 29, 125
fiber cladding 23
fiber laser 15
focus shift 27
focusing 25, 28
forming 60
fresnel absorption 38, 67, 106
Fresnel 38

G

gaps 75, 84, 117, 146
Gaussian beam 6, 19, 30, 98
gradient mechanism 61

H

hardening 32, 54, 115
hatch distance 117, 136
heat-affected zone (haz) 66, 86, 103, 110, 113
heat conduction 34, 40
heat conduction welding 66
hermite-gaussian beam 7
humping 99
hybrid welding 92

I

incomplete penetration 100
intensity 34, 37, 51, 67, 71, 106, 127
inverse bremsstrahlung 76

J

joining 64, 66

K

keyhole 66, 71, 74
Knudsen layer 39, 77

L

Laguerre-gaussian beam 7
laser active medium 3, 8
laser classes 53
laser machine 3
lenses 26
line energy 52

M

Marangoni flow 43, 69, 77, 130
marking 62
martensite 55
masking 63
melt pool 39, 43, 52, 68, 78, 96, 102, 110, 116, 124
microstructure 44, 66
mirrors 25
monitoring 46, 82, 131, 137
monitoring 46, 79, 82, 131, 137, 147
monochromatic 19
moving point source 41
moving line source 41
multiple reflections 16, 34, 54, 75
multi-mode 31
mushy zone 104

N

neodymium (nd) 5, 7, 13, 66
nitrogen (N_2) 9
numerical aperture 23

O

oscillation optics 31, 63

P

parabolic mirror 15
peening 60

phase changes 43
photon 2, 4, 8, 12, 19, 112
polarization 36, 38, 108, 114
polarizor 26
population inversion 4, 11
porosity 101, 126
powder 118, 124, 126, 127, 133
powder Bed Fusion 133, 144
power 13
power density 2
pressure balance 74
process emission 46
pulsed lasers 16
pumping 3, 11

Q

quality 94, 110, 116, 131
quantum efficiency 8

R

Rayleigh length 20, 29
recoil pressure 74, 109, 113, 140
recycling 137
reflection 2
refractive index 15, 23
ripples 110
relaxation 10
remote processing 31, 51
repair 116, 123, 124
resonator 3, 5, 16
rod laser 13

S

safety 53, 87, 134
scanning optics 31
sensors 47
shielding gas 64, 68, 80, 91, 93, 97, 101, 103, 118, 124, 139
slicing 123, 133
solid state lasers 11
spatial intensity distribution 24, 31, 117
spatter 96, 131, 137
speed of light 2
spherical aberration 27
spontaneous emission 3
stable resonator 6
standing wave 6
stimulated emission 3
support structures 133
surface hardening 54
surface tension 43, 65, 69, 74, 79, 96, 99, 108, 130, 145

T

thermal cycle 44, 56, 67, 108, 137
thermal deformation 61, 131
thermal efficiency 35
thermal expansion 14, 27
thermal gradient 26, 61, 70,
    105, 113
thermal lensing 15, 26
thermal radiation 35, 46
thermal stress 104, 113, 131
top-hat beam 24, 30
total reflection 23
transmission 2
triangulation 49

U

undercuts 99
unstable resonator 6

V

vapor channel 71, 77, 78, 96, 146
vaporization 52, 73, 77, 79, 81, 111, 115, 125, 139
vapor plume 39, 77, 139, 145
vibration states 9

W

wall plug efficiency 8
wave front 19
wavelength 2, 19
weldability 85
welding 51, 66
wetting 64

Y

Ytterbium (Yb) 7, 14, 15
Yttrium-Aluminum-Garnet (YAG) 5, 13, 14, 105

Printed in the United States
by Baker & Taylor Publisher Services